T0073112

Generative Adversarial Networks for Image Generation

Xudong Mao • Qing Li

Generative Adversarial Networks for Image Generation

Xudong Mao
Department of Computing
Hong Kong Polytechnic University
Hong Kong, China

Qing Li
Department of Computing
Hong Kong Polytechnic University
Hong Kong, China

ISBN 978-981-33-6047-1 ISBN 978-981-33-6048-8 (eBook)
https://doi.org/10.1007/978-981-33-6048-8

This Springer imprint is published by the registered company Springer Nature Singapore Pte Ltd.
The registered company address is: 152 Beach Road, #21-01/04 Gateway East, Singapore 189721, Singapore

Preface

Image generation is a typical application of unsupervised learning, it is implemented by learning the distribution over images and generating new images by sampling from the learned distribution. It is an important and fundamental task in computer vision, and the recent success of image generation achieved by deep generative models has driven the progress of numerous computer vision tasks. In this book we introduce the methods based on Generative Adversarial Networks (GANs) for image generation. Regular GANs hypothesize the discriminator as a classifier with the sigmoid cross entropy loss function. However, the sigmoid cross entropy loss may lead to the vanishing gradients problem during the learning process. To address this issue, we propose the Least Squares (LSGANs), which adopt the least squares loss for both the discriminator and the generator.

Multi-domain image generation is an extension of image generation, which aims to generate aligned image pairs of different domains. It also has many promising applications such as image-to-image translation and unsupervised domain adaptation. We introduce a model called Regularized Conditional GANs (RCGANs) for multi-domain image generation. RCGANs are based on the conditional GANs, and two regularizers are used to guide the model to encode the common semantics in the shared latent variables and encode the domain-specific semantics in the domain variables.

Although the main targeted application of this book is image generation, the developed theories and techniques are applicable to other types of applications. In Chap. 3, we introduce three such interesting applications of GANs, including image-to-image translation, unsupervised domain adaptation, and GANs for security.

This book is designed for researchers and practitioners in the fields of computer vision and machine learning. We expect the reader to have some basic knowledge in these fields, including upper-division undergraduate students and postgraduate (master's and PhD) students in computer science, computer engineering, and/or electrical and information engineering.

Hong Kong, China
October 2020

Xudong Mao
Qing Li

Acknowledgments

Dr. Celine Chang was the first editor from Springer Nature who encouraged us enthusiastically to come up with this timely book. Ms. Saanthi Shankhararaman as the project coordinator (book) has been patiently yet persistently reminding us from time to time on the progress of developing this book on time. We also acknowledge the great help from AsiaEdit in Hong Kong, for their careful and professional editing services rendered to us.

We owe our heart-felt gratitude to the members of our research group who have offered many helps to the writing of the book: Yun Ma, Yangbin Chen, Zhenguo Yang, Min Cheng, Wenqi Fan, Aurele T. Gnanha, Yaowei Wang, Xueying Zhan, Xiao Yu, Xiaoye Miao, Zongxi Li, Xinhong Chen, Runze Mao, and Xiaoya Chong; in addition, we also thank Prof. Zhiguo Gong, Prof. Jianming Lv, Prof. Haiwei Pan, Dr. Xiaoye Miao, and Prof. Xiande Wu for their valuable inputs, sincere friendships, and tremendous help.

We would like to thank our families. This project took valuable time away from them during the last 12 months. We appreciate their kind understanding and patience, as well as their great support during the writing of this book.

October 2020

Xudong Mao
Qing Li

Contents

Acronyms

This section provides a reference describing the abbreviations used throughout this book.

ADDA	Adversarial Discriminative Domain Adaptation
AMT	Amazon Mechanical Turk
BN	Batch Normalization
CAN	Collaborative and Adversarial Network
CDAN	Conditional Adversarial Domain Adaptation
cGAN	Conditional Generative Adversarial Network
CNN	Convolutional Neural Network
CNN-RNN	Convolutional Recurrent Neural Network
CoGAN	Coupled Generative Adversarial Network
CUT	Contrastive Unpaired Translation
CycleGAN	Cycle-Consistent Adversarial Network
DANN	Domain Adversarial Neural Network
DA-GAN	Deep Attention Generative Adversarial Network
DBM	Deep Boltzmann Machine
DBN	Deep Belief Network
DCGAN	Deep Convolutional Generative Adversarial Network
DIRT	Decision-Boundary Iterative Refinement Training
DRIT	Disentangled Representation for Image-to-Image Translation
DVP	Deep Video Portraits
FID	Fréchet Inception Distance
FSGAN	Face Swapping Generative Adversarial Network
GAN	Generative Adversarial Network
IS	Inception Score
LAPGAN	Laplacian Pyramid of Generative Adversarial Network
LRCN	Long-Term Recurrent Convolutional Neural Network
LSGAN	Least Squares Generative Adversarial Network
LSGAN-GP	Least Squares Generative Adversarial Network with Gradient Penalty

LSTM	Long Short-Term Memory
M-GAN	Minimax Generative Adversarial Network
NADE	Neural Autoregressive Distribution Estimator
NS-GAN	Non-saturating Generative Adversarial Network
paGAN	Photoreal Avatar Generative Adversarial Network
PAN	Perceptual Adversarial Network
RBM	Restricted Boltzmann Machine
RCGAN	Regularized Conditional Generative Adversarial Network
RNN	Recurrent Neural Network
SPADE	Spatially Adaptive Normalization
SVM	Support Vector Machine
UNIT	Unsupervised Image-to-Image Translation
VADA	Virtual Adversarial Domain Adaptation
VAE	Variational Autoencoder
WGAN	Wasserstein Generative Adversarial Network
WGAN-GP	Wasserstein Generative Adversarial Network with Gradient Penalty

Chapter 1
Generative Adversarial Networks (GANs)

1.1 Introduction to GANs

Deep learning has launched a profound reformation and has even been applied to many real-world tasks such as image classification (He et al. 2016), object detection (Ren et al. 2015), and image segmentation (Long et al. 2015). These tasks all fall into the scope of supervised learning, which means that large amounts of labeled data are provided for the learning processes. Compared with supervised learning, however, unsupervised learning shows little effect from deep learning. Generative modeling is a typical problem in unsupervised learning, the goal of which is to learn the distribution over training data and then to generate new data by sampling from the learned distribution. Generative modeling is usually more difficult than supervised learning tasks because the learning criteria of generative modeling are intractable (Goodfellow et al. 2016). For supervised learning tasks, the corresponding mapping information between the inputs and the outputs is given, and the supervised learning models need only learn how to encode the mapping information into the neural networks. In contrast, for generative modeling, the correspondence between the inputs (usually a noise vector) and the outputs (the training data) is unknown, and the generative models must learn how to arrange the mapping between the inputs and the outputs efficiently.

Generative models can be classified into two categories, undirected generative models and directed generative models, based on whether the interactions between neural layers are directed or undirected, as shown in Fig. 1.1. Undirected generative models include Restricted Boltzmann Machine (RBM) (Hinton and Salakhutdinov 2006) and Deep Boltzmann Machine (DBM) (Salakhutdinov and Hinton 2009). Undirected generative models usually have the problem of the intractable partition function, and the techniques used to approximate it limit their effectiveness. Recently, directed generative models have been popular, and many powerful models have been proposed, such as Neural Autoregressive Distribution Estimator (NADE) (Larochelle and Murray 2011) and Variational Autoencoder

X. Mao, Q. Li, *Generative Adversarial Networks for Image Generation*,
https://doi.org/10.1007/978-981-33-6048-8_1

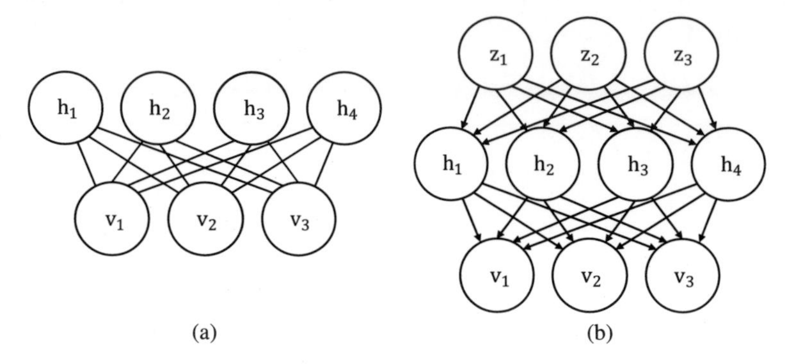

Fig. 1.1 Examples of undirected and directed generative models. (**a**) Undirected generative model. (**b**) Directed generative model

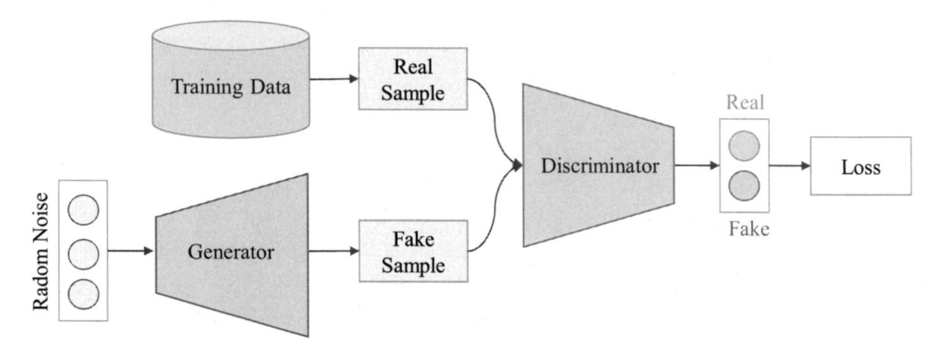

Fig. 1.2 Framework of GAN

(VAE) (Kingma and Welling 2013). Examples of undirected and directed generative models are shown in Fig. 1.1.

Directed generative models usually contain a feedforward network that transforms the latent variables $\mathbf{z}$ to the observed samples $\mathbf{x}$, as shown in Fig. 1.1b. The latent variables $\mathbf{z}$ are usually sampled from a simple distribution (e.g., Gaussian distribution). We can think of the idea of directed generative models as to map a simple distribution over $\mathbf{z}$ to a complex (observed) distribution over $\mathbf{x}$ via a feedforward network.

Generative Adversarial Networks (GANs) (Goodfellow et al. 2014) are one type of directed generative model. GANs can generate very realistic images. The framework of GANs is shown in Fig. 1.2. GANs consist of two networks: the generator and the discriminator. The generator is the typical feedforward network in a directed generative model that maps the latent variables $\mathbf{z}$ to the observed samples $\mathbf{x}$. The discriminator is a classifier that distinguishes between real samples and fake samples generated by the generator. The generator and the discriminator are trained adversarially: the discriminator aims to distinguish between real samples and fake

Table 1.1 The input and output of the discriminator and the generator

	Discriminator	Generator
Input	A real image from training data or a fake image from the generator	A random noise vector
Output	Probability that the input is a real image	A fake image
Loss	Classify real images into the "real" category and fake images into the "fake" category	Classify fake images into the "real" category (from the discriminator)

samples, and the generator tries to generate fake samples as real as possible to make the discriminator believe that the fake samples are from real data.

GANs have been one of the most successful generative models and have even served as the basis for many computer vision and graphics tasks. If the task is to output high-quality images, GANs can help improve the image quality. The general idea is to add a discriminator and incorporate the GANs loss into the objective of the task. GANs have been applied to many specific tasks, such as image super-resolution (Ledig et al. 2017), text to image synthesis (Reed et al. 2016), and image-to-image translation (Isola et al. 2017).

Table 1.1 lists the input and output of the discriminator and the generator. The discriminator is a typical classifier that predicts whether the input image is real or fake (i.e., derived from the training data or from the generator). The generator starts from a random vector and maps the vector to an image (typically through some transposed convolutional layers). Note that the loss of the generator is from the discriminator, such that the generator can know how the discriminator behaves. If the discriminator can distinguish a fake image correctly, the generator tries to improve that image. If the generator can generate a higher-quality fake image, the discriminator becomes better able to distinguish that fake image, and a better discriminator allows the generator to further improve. Therefore, the generator can finally generate very realistic images.

The details of GANs training are shown in Algorithm 1. The generator and the discriminator are updated alternately. When updating the generator, the parameters of the discriminator are fixed, and vice versa. The objective of the discriminator is to minimize the classification error (e.g., the binary cross-entropy loss). For the generator, it aims to fool the discriminator into making false predictions for the fake images. Thus, we can set the objective of the generator to maximize the classification error of the discriminator.

1.2 Challenges of GANs

Although GANs have achieved great success, they still have three main challenges. The first challenge is image quality. Many studies have sought to improve the image quality of GANs (Radford et al. 2015; Denton et al. 2015; Huang et al. 2017; Salimans et al. 2016; Odena et al. 2016; Arora et al. 2017; Berthelot et al. 2017; Zhang et al. 2018; Brock et al. 2018; Jolicoeur-Martineau 2018). Radford et al. (2015) proposed the Deep Convolutional GANs (DCGANs), which are the first to successfully introduce convolutional layers to GANs architectures. Denton et al. (2015) proposed the Laplacian pyramid of GANs (LAPGANs), in which a Laplacian pyramid is constructed to generate high-resolution images based on low-resolution images. Karras et al. (2017) proposed a new training method called progressive training, which first generates realistic images at a resolution of 1024×1024. Brock et al. (2018) pointed out that GANs benefit dramatically from scaling up the batch size and the number of channels in each layer, and the proposed BigGANs substantially improve the performance on complex datasets such as ImageNet (Russakovsky et al. 2015). Although GANs can generate photorealistic images that even humans cannot distinguish from real images, it is limited to some simple datasets whose objects are highly "templated" and centered with small margins, such as face datasets. For complex datasets, such as scene datasets, the performance of GANs remains limited, and people can easily distinguish the generated images from real ones.

The second challenge is training stability. Generally speaking, training GANs is difficult in practice due to the mode collapse problem (Radford et al. 2015; Metz et al. 2016). Mode collapse means that the generator can fool the discriminator by generating from only one mode (i.e., generating very similar images). Some works (Arjovsky et al. 2017; Nowozin et al. 2016) sought to address this problem by analyzing the objective functions of GANs. Nowozin et al. (2016) proposed the f-GANs that generalize the original GANs (Goodfellow et al. 2014), which correspond to Jensen-Shannon divergence, to any type of f-divergence. Arjovsky et al. (2017) proposed the Wasserstein GANs (WGANs), which use the Wasserstein distance to measure the distance between the generated and real samples. Regularization techniques are also effective in improving the training stability of GANs, such as gradient penalty (Gulrajani et al. 2017) and spectral normalization (Miyato et al. 2018). Gulrajani et al. (2017) proposed that a penalty on the gradient norm can be used to enforce the Lipschitz constraint in the Wasserstein distance. Miyato et al. (2018) proposed spectral normalization that constrains the spectral norm of each layer to control the Lipschitz constant of the discriminator. Note that the improvement of training stability can usually lead to higher-quality generated images.

The third challenge is the evaluation of GANs. Inception score (IS) (Salimans et al. 2016) and Fréchet inception distance (FID) (Heusel et al. 2017) are two widely used evaluation metrics for GANs. IS correlates the image quality with the degree to which the images are highly classifiable using a pre-trained classifier.

Algorithm 1 Training process of GANs

for number of training iterations **do**

- Sample a batch of real images $\boldsymbol{x}$ from training data.
- Sample a batch of noise vectors z from Gaussian distribution.
- Use z to generate a batch of fake images $\boldsymbol{x}^*$ from the generator.
- Update the discriminator using UPDATE_DISCRIMINATOR($\boldsymbol{x}, \boldsymbol{x}^*$).
- Update the generator using UPDATE_GENERATOR($\boldsymbol{x}^*$).

end for

function UPDATE_DISCRIMINATOR($\boldsymbol{x}, \boldsymbol{x}^*$)

- Compute the discriminator's prediction for $\boldsymbol{x}$ and $\boldsymbol{x}^*$.
- Compute the classification error for $\boldsymbol{x}$ and $\boldsymbol{x}^*$.
- Update the discriminator's parameters to minimize the classification error.

end function

function UPDATE_GENERATOR($\boldsymbol{x}^*$)

- Compute the discriminator's prediction for $\boldsymbol{x}^*$.
- Compute the classification error for $\boldsymbol{x}^*$.
- Update the generator's parameters to maximize the classification error of the discriminator.

end function

FID models the features of generated and real data as continuous multivariate Gaussian distributions and uses the Fréchet distance to measure the distance between generated and real data. Although IS and FID are widely used, questions remain, such as the use of pre-trained networks and the approximations of Gaussian distributions (Borji 2019).

Bibliography

Arjovsky M, Chintala S, Bottou L (2017) Wasserstein GAN. In: International conference on machine learning (ICML), pp 214–223

Arora S, Ge R, Liang Y, Ma T, Zhang Y (2017) Generalization and equilibrium in generative adversarial nets (GANs). arXiv:1703.00573

Berthelot D, Schumm T, Metz L (2017) BEGAN: boundary equilibrium generative adversarial networks. arXiv:1703.10717

Borji A (2019) Pros and cons of GAN evaluation measures. Comput Vis Image Underst 179:41–65

Brock A, Donahue J, Simonyan K (2018) Large scale GAN training for high fidelity natural image synthesis. arXiv:1809.11096

Denton E, Chintala S, Szlam A, Fergus R (2015) Deep generative image models using a Laplacian pyramid of adversarial networks. In: Advances in neural information processing systems (NeurIPS), pp 1486–1494

Goodfellow I, Pouget-Abadie J, Mirza M, Xu B, Warde-Farley D, Ozair S, Courville A, Bengio Y (2014) Generative adversarial nets. In: Advances in neural information processing systems (NeurIPS), pp 2672–2680

Goodfellow I, Bengio Y, Courville A (2016) Deep learning. MIT Press, Cambridge, MA

Gulrajani I, Ahmed F, Arjovsky M, Dumoulin V, Courville A (2017) Improved training of Wasserstein GANs. In: Advances in neural information processing systems (NeurIPS), pp 5767–5777

He K, Zhang X, Ren S, Sun J (2016) Deep residual learning for image recognition. In: Computer vision and pattern recognition (CVPR), pp 770–778

Heusel M, Ramsauer H, Unterthiner T, Nessler B, Hochreiter S (2017) GANs trained by a two time-scale update rule converge to a local nash equilibrium. In: Advances in neural information processing systems (NeurIPS), pp 6626–6637

Hinton G, Salakhutdinov R (2006) Reducing the dimensionality of data with neural networks. Science 313(5786):504–507

Huang X, Li Y, Poursaeed O, Hopcroft J, Belongie S (2017) Stacked generative adversarial networks. In: Computer vision and pattern recognition (CVPR), pp 5077–5086

Isola P, Zhu J-Y, Zhou T, Efros AA (2017) Image-to-image translation with conditional adversarial networks. In: Computer vision and pattern recognition (CVPR), pp 5967–5976

Jolicoeur-Martineau A (2018) The relativistic discriminator: a key element missing from standard GAN. arXiv:1807.00734

Karras T, Aila T, Laine S, Lehtinen J (2017) Progressive growing of GANs for improved quality, stability, and variation. arXiv:1710.10196

Kingma DP, Welling M (2013) Auto-encoding variational Bayes. arXiv:1312.6114

Larochelle H, Murray I (2011) The neural autoregressive distribution estimator. In: International conference on artificial intelligence and statistics (AISTATS), pp 29–37

Ledig C, Theis L, Huszar F, Caballero J, Cunningham A, Acosta A, Aitken A, Tejani A, Totz J, Wang Z, Shi W (2017) Photo-realistic single image super-resolution using a generative adversarial network. In: Computer vision and pattern recognition (CVPR), pp 4681–4690

Long J, Shelhamer E, Darrell T (2015) Fully convolutional networks for semantic segmentation. In: Computer vision and pattern recognition (CVPR), pp 3431–3440

Metz L, Poole B, Pfau D, Sohl-Dickstein J (2016) Unrolled generative adversarial networks. arXiv:1611.02163

Miyato T, Kataoka T, Koyama M, Yoshida Y (2018) Spectral normalization for generative adversarial networks. arXiv:1802.05957

Nowozin S, Cseke B, Tomioka R (2016) f-GAN: training generative neural samplers using variational divergence minimization. In: Advances in neural information processing systems (NeurIPS), pp 271–279

Odena A, Olah C, Shlens J (2016) Conditional image synthesis with auxiliary classifier GANs. arXiv:1610.09585

Radford A, Metz L, Chintala S (2015) Unsupervised representation learning with deep convolutional generative adversarial networks. arXiv:1511.06434

Reed S, Akata Z, Yan X, Logeswaran L, Schiele B, Lee H (2016) Generative adversarial text-to-image synthesis. In: International conference on machine learning (ICML), pp 1060–1069

Ren S, He K, Girshick R, Sun J (2015) Faster R-CNN: towards real-time object detection with region proposal networks. In: Advances in neural information processing systems (NeurIPS), pp 91–99

Russakovsky O, Deng J, Su H, Krause J, Satheesh S, Ma S, Huang Z, Karpathy A, Khosla A, Bernstein M, Berg AC, Fei-Fei L (2015) ImageNet large scale visual recognition challenge. Int J Comput Vis 115:211–252

Salakhutdinov R, Hinton G (2009) Deep Boltzmann machines. In: International conference on artificial intelligence and statistics, pp 448–455

Salimans T, Goodfellow I, Zaremba W, Cheung V, Radford A, Chen X, Chen X (2016) Improved techniques for training GANs. In: Advances in neural information processing systems (NeurIPS), pp 2226–2234

Zhang H, Goodfellow I, Metaxas D, Odena A (2018) Self-attention generative adversarial networks. arXiv:1805.08318

Chapter 2
GANs for Image Generation

2.1 Image Generation

2.1.1 Overview of Image Generation

Deep learning has proven to be hugely successful in computer vision and has even been applied to many real-world tasks, such as image classification (He et al. 2016), object detection (Ren et al. 2015), and segmentation (Long et al. 2015). Compared with these tasks in supervised learning, however, image generation, which belongs to unsupervised learning, may not achieve the desired performance. The target of image generation is to learn to draw pictures by means of some generative models, as shown in Fig. 2.1.

Why would image generation be difficult? For supervised learning tasks, the corresponding mapping information between the inputs and the outputs is given, and the supervised learning models need only learn how to encode the mapping information into the neural networks. In contrast, for image generation, the correspondence between the input and the output is unknown, and the generative models must learn how to arrange the mapping between the inputs and the outputs efficiently. Another reason is the huge output dimension of image generation. For supervised learning, it contains a limited number of output dimensions. For example, the output dimension for ImageNet classification is 1000. For image generation, however, if the image resolution is 1024×1024, the output dimension will be larger than three million. This requires the model to learn the features efficiently.

Why would image generation be useful? Image generation is an important and fundamental problem in computer vision, and progress in this area can boost the performance of numerous computer vision tasks. In general, if a model for image generation can generate high-quality images, it can be used to improve the performance of tasks that involve the generation of high-quality images such as image super-resolution (Ledig et al. 2017), image-to-image translation (Isola et al. 2017), and image compression (Agustsson et al. 2018).

X. Mao, Q. Li, *Generative Adversarial Networks for Image Generation*,
https://doi.org/10.1007/978-981-33-6048-8_2

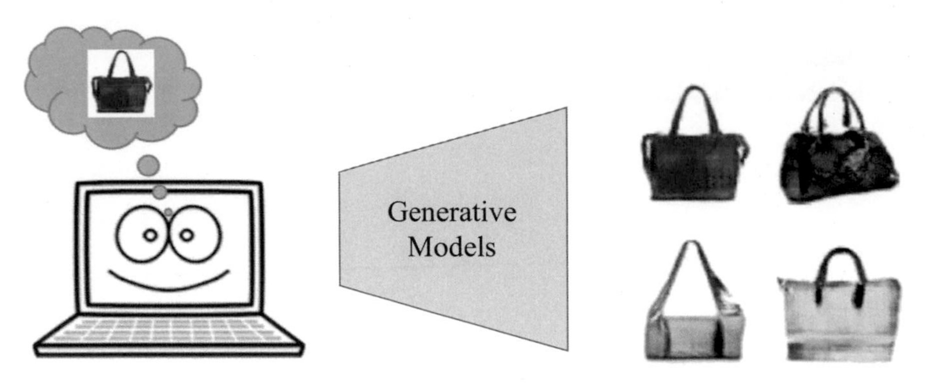

Fig. 2.1 Illustration of image generation

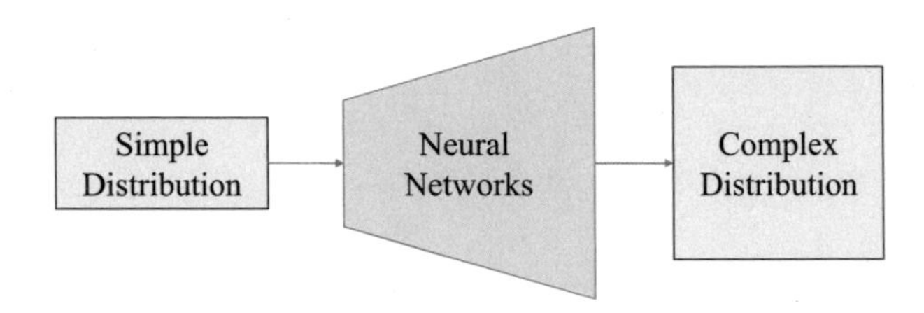

Fig. 2.2 Illustration of deep directed generative models

Many deep directed generative models have been proposed to address the problem of image generation. Deep directed generative models first sample from a simple distribution (e.g., Gaussian distribution) and then use neural networks to learn a transformation to map the simple distribution to a complex distribution, as shown in Fig. 2.2. Modern deep directed generative models include VAEs (Kingma and Welling 2013), flow-based models (Dinh et al. 2014), and GANs (Goodfellow et al. 2014).

Recently, GANs have proven their effectiveness in image generation. The idea of GANs is to train a discriminator and a generator in an adversarial manner. The discriminator aims to distinguish between real samples and fake samples; on the contrary, the generator tries to learn the transformation to generate fake samples that look as real as possible.

Although GANs have achieved great success, three main challenges remain, as described in Sect. 1.2. The first challenge is image quality. Numerous works have been proposed to improve the quality of generated images, such as DCGANs (Radford et al. 2015) and LAPGANs (Denton et al. 2015). The second challenge is training stability. Training GANs is a difficult issue in practice due to the mode collapse problem (Radford et al. 2015; Metz et al. 2016). The third challenge is the evaluation of GANs.

In this book, we focus on the challenges of image quality and training stability, and investigate several approaches to address these two challenges. We first argue that the objective function of regular GANs may lead to the vanishing gradient problem. To overcome such a problem, we propose to use the least squares loss for both the discriminator and the generator. We show that our proposed model can improve both the image quality and training stability of GANs. We then propose two parameter schemes for the least squares loss. One is derived from theoretical analysis, which yields minimizing the Pearson χ^2 divergence. The other scheme is to follow the traditional way of using least squares loss for classification. We also explore the use of the gradient penalty to further improve training stability.

We then study another task called multi-domain image generation, which is an extension of image generation. The goal of multi-domain image generation is to generate pairs of corresponding images from different domains. Multi-domain image generation is an important extension of image generation. It has many promising applications such as improving the generated image quality (Dosovitskiy et al. 2015; Wang and Gupta 2016), image-to-image translation (Perarnau et al. 2016; Wang et al. 2017a), and unsupervised domain adaptation (Liu and Tuzel 2016). The difficulty of multi-domain image generation is to learn the correspondence of different domain images, especially when the information of paired samples is not given. To tackle this problem, we propose a model based on conditional GANs that can learn to generate corresponding images in the absence of paired training data.

2.1.2 *GANs for Image Generation*

Image generation is an important and fundamental problem in computer vision, the goal of which is to learn the distribution over images and then to generate images that look as real as possible by sampling from the learned distribution. Recently, deep learning has achieved great success in image generation, and many successful deep generative models have been proposed, such as RBM (Hinton and Salakhutdinov 2006), DBM (Salakhutdinov and Hinton 2009), and VAE (Kingma and Welling 2013). However, these models all face the difficulties of intractable functions (e.g., intractable partition function) or intractable inference, which in turn restricts their effectiveness.

Unlike the above deep generative models, which usually adopt approximation methods for intractable functions or inference, GANs (Goodfellow et al. 2014) require no approximate inference and can be trained end-to-end via a differentiable network (Goodfellow et al. 2016). The basic idea of GANs is to train a discriminator and a generator simultaneously: the discriminator aims to distinguish between real samples and generated samples, while the generator tries to generate fake samples that look as real as possible to make the discriminator believe that the fake samples are from real data. GANs have demonstrated impressive performance for various computer vision tasks such as image generation (Nguyen et al. 2017; Chen et al. 2016), image super-resolution (Ledig et al. 2017), and semi-supervised learning (Springenberg 2018; Salimans et al. 2016).

Formally, the target of the generator G is to learn the distribution p_g over data $\boldsymbol{x}$. G starts with sampling input variables z from a uniform or Gaussian distribution $p_z(z)$ and then maps the input variables z to data space $G(z; \theta_g)$ via a differentiable network. In contrast, the discriminator D is a classifier $D(\boldsymbol{x}; \theta_d)$ that aims to recognize whether an image is from training data or from G. The original GANs paper (Goodfellow et al. 2014) adopted the sigmoid cross-entropy loss for the discriminator:

$$\min_G \max_D V_{\text{GAN}}(D, G) = \mathbb{E}_{\boldsymbol{x} \sim p_{\text{data}}(\boldsymbol{x})}[\log D(\boldsymbol{x})] + \mathbb{E}_{z \sim p_z(z)}[\log(1 - D(G(z)))]. \tag{2.1}$$

For the generator, they presented two different losses for the generator: the "minimax" loss (M-GANs) and the "non-saturating" loss (NS-GANs):

$$\mathbf{M-GAN}: \quad \min_G V_{\text{GAN}}(G) = \mathbb{E}_{z \sim p_z(z)}[\log(1 - D(G(z)))], \tag{2.2}$$

$$\mathbf{NS-GAN}: \quad \min_G V_{\text{GAN}}(G) = -\mathbb{E}_{z \sim p_z(z)}[\log(D(G(z)))]. \tag{2.3}$$

They pointed out that M-GANs will saturate at the early stage of the learning process. Thus, NS-GANs are recommended for use in practice.

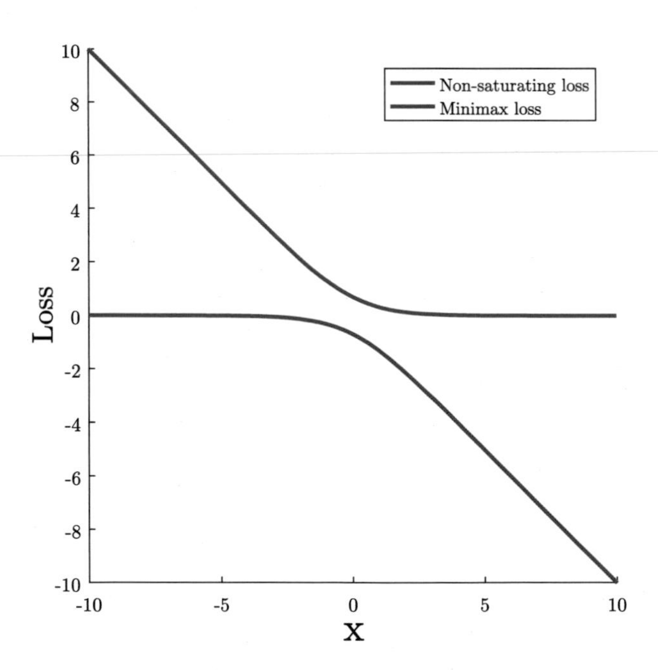

Fig. 2.3 Non-saturating loss and minimax loss

We argue that both the non-saturating loss and the minimax loss will lead to the problem of vanishing gradients when updating the generator. As Fig. 2.3 shows, the non-saturating loss will saturate when the input is relatively large, whereas the minimax loss will saturate when the input is relatively small. Consequently, as Fig. 2.4c shows, when updating the generator, the non-saturating loss will cause almost no gradient for the fake samples in magenta, because these samples are on the side of real data, corresponding to the input with relatively large values. Similarly, the minimax loss will cause almost no gradient for the fake samples in green. However, these fake samples are still far from real data, and we wish to pull them closer to real data. Based on this observation, we propose Least Squares GANs (LSGANs), which adopt the least squares loss for both the discriminator and the generator. The idea is simple yet powerful: the least squares loss can move the fake samples toward the decision boundary, because the least squares loss penalizes samples that lie far from the decision boundary even though they are on the correct side. As Fig. 2.4d shows, the least squares loss will penalize those two types of fake samples and pull them toward the decision boundary. Based on this property, LSGANs can generate samples that are closer to real data.

Another benefit of LSGANs is improved training stability. Generally speaking, training GANs is difficult in practice because of the instability of GANs learning (Radford et al. 2015; Metz et al. 2016; Arjovsky and Bottou 2017). Recently, several papers have noted that the instability of GANs learning is partially caused by the objective function (Arjovsky et al. 2017; Metz et al. 2016; Qi 2017). Specifically, minimizing the objective function of regular GANs may cause the problem of vanishing gradients, which makes it difficult to update the generator.

Our proposed LSGANs can alleviate the mode collapse problem because the least squares loss penalizes samples based on the distance to the decision boundary, which in turn generates more gradients when updating the generator. Moreover, we theoretically show that the training instability of regular GANs is a result of the mode-seeking behavior (Bishop 2006) of the objective function, whereas LSGANs exhibit less mode-seeking behavior.

We further improve the training stability of LSGANs by the use of gradient penalty (Kodali et al. 2017), because gradient penalty has shown the effectiveness of improving the training stability of GANs (Kodali et al. 2017; Gulrajani et al. 2017). We find that LSGANs with gradient penalty (LSGANs-GP) can train successfully for all the difficult architectures used in WGANs-GP (Gulrajani et al. 2017). However, gradient penalty also has some inevitable disadvantages such as additional computational cost and memory cost. Based on this observation, we evaluate the stability of LSGANs in two settings: LSGANs without gradient penalty and LSGANs with gradient penalty.

Evaluation of the training stability of GANs remains an open issue. One popular evaluation method is to use difficult architectures, such as by excluding the batch normalization (Arjovsky et al. 2017). However, in practice, stable architectures will always be selected for tasks. Sometimes the difficulty stems from the datasets. Motivated by this, we propose to use difficult datasets but stable architectures to evaluate the stability of GANs. Specifically, we create two synthetic digit datasets

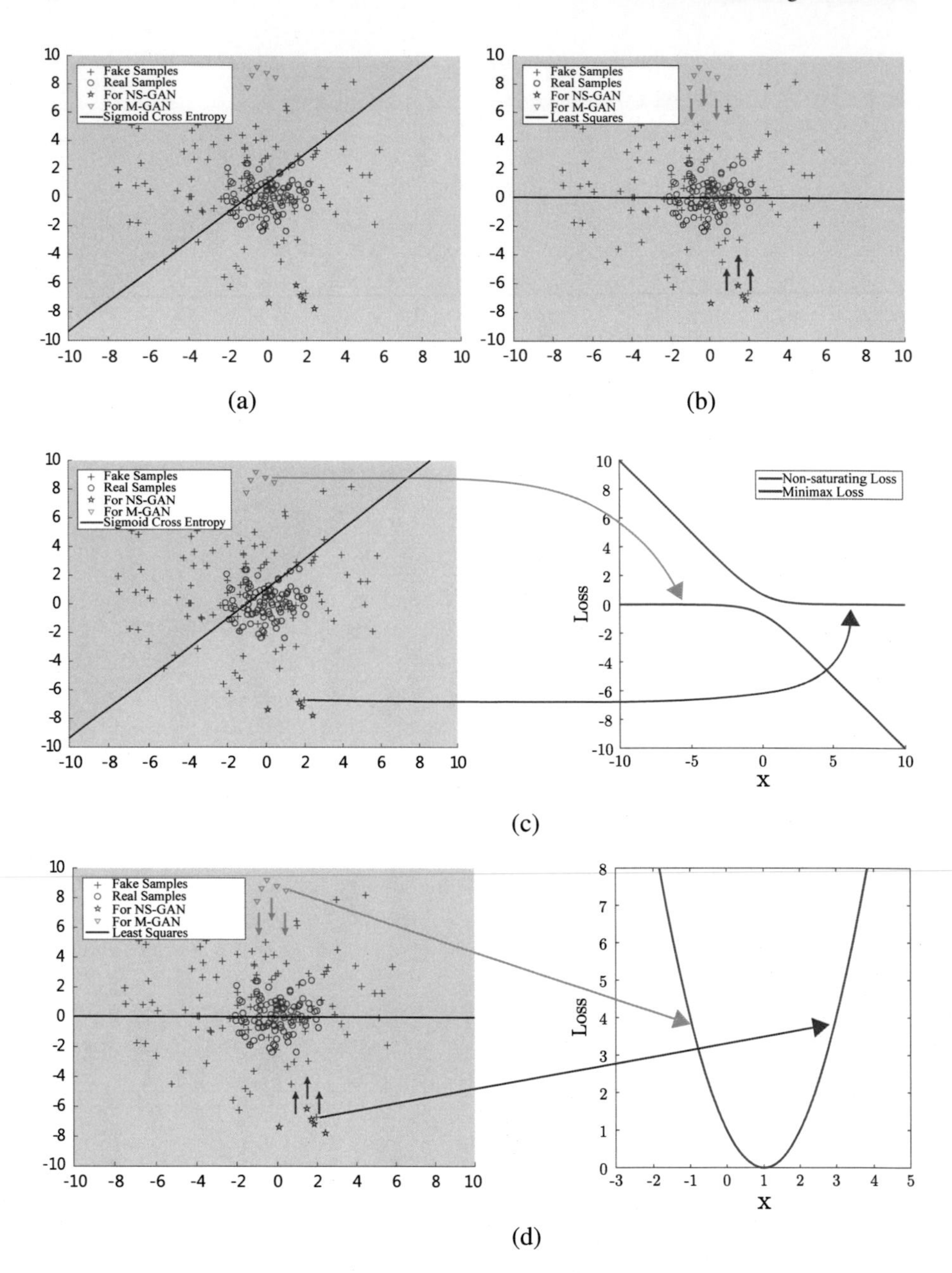

Fig. 2.4 Illustration of different behaviors of two loss functions. (**a**): Decision boundary of the sigmoid cross-entropy loss function. The orange area is the side of the real samples, and the blue area is the side of the fake samples. The non-saturating loss and the minimax loss will cause almost no gradient for the fake samples in magenta and green, respectively, when used to update the generator. (**b**): Decision boundary of the least squares loss function. It penalizes the fake samples (both in magenta and green), and as a result, it forces the generator to generate samples toward the decision boundary. (**c**): Behavior of the non-saturating loss and the minimax loss. (**d**): Behavior of the least squares loss

with small variability by rendering 28×28 digits using some standard fonts. Datasets with small variability are difficult for GANs to learn, because the discriminator can distinguish the real samples very easily for such datasets.

2.1.3 Background Research of GANs

Deep generative models attempt to capture the probability distributions over the given data. Restricted Boltzmann Machines (RBMs), one type of deep generative model, form the basis of many other hierarchical models, and they have been used to model the distributions of images (Taylor et al. 2010) and documents (Hinton and Salakhutdinov 2009). Deep Belief Networks (DBNs) (Hinton et al. 2006) and Deep Boltzmann Machines (DBMs) (Salakhutdinov and Hinton 2009) are extended from the RBMs. The most successful application of DBNs is image classification (Hinton et al. 2006), where DBNs are used to extract feature representations. However, RBMs, DBNs, and DBMs all have the difficulties of intractable partition functions or intractable posterior distributions, which thus require the use of approximation methods to learn the models. Another important deep generative model is Variational Autoencoders (VAE) (Kingma and Welling 2013), a directed model, which can be trained with gradient-based optimization methods. However, VAEs are trained by maximizing the variational lower bound, which may lead to the blurriness problem of generated images (Goodfellow et al. 2016).

Recently, GANs have been proposed by Goodfellow et al. (2014), who explained the theory of GANs learning based on a game-theoretic scenario. A similar idea was also introduced by Ganin et al. (2016), where a method of adversarial training was proposed for domain adaptation. Showing the powerful capability for unsupervised tasks, GANs have been applied to many specific tasks, like data augmentation (Shrivastava et al. 2016), image super-resolution (Ledig et al. 2017), text-to-image synthesis (Reed et al. 2016), video prediction (Mathieu et al. 2015; Vondrick et al. 2016), image compression (Agustsson et al. 2018), visual manipulation (Yan et al. 2015; Zhu et al. 2016), and image-to-image translation (Taigman et al. 2016; Isola et al. 2017; Wang et al. 2017b, 2018). By combining traditional content loss and adversarial loss, super-resolution GANs (Ledig et al. 2017) have achieved state-of-the-art performance for the task of image super-resolution. Reed et al. (2016) proposed a model to synthesize images given text descriptions based on the conditional GANs (Mirza and Osindero 2014). Isola et al. (2017) also used conditional GANs to transfer images from one representation to another. In addition to unsupervised learning tasks, GANs also show good potential for semi-supervised learning tasks. Salimans et al. (2016) proposed a GAN-based framework for semi-supervised learning, in which the discriminator not only outputs the probability that an input image is from real data; it also outputs the probabilities of belonging to each class. Another important problem of GANs is to infer the latent vectors from the given examples (Larsen et al. 2015; Makhzani et al. 2015; Donahue et al. 2016; Dumoulin et al. 2016; Li et al. 2017; Mescheder et al. 2017b). Both Donahue et al.

(2016) and Dumoulin et al. (2016) proposed a bidirectional adversarial learning framework by incorporating an encoder into the GANs framework. Li et al. (2017) proposed to use the conditional entropy to regularize their objectives (Donahue et al. 2016; Dumoulin et al. 2016), making the learning process more stable.

Despite the great successes GANs have achieved, improving the quality of generated images remains a challenge. Many studies have been proposed to improve the quality of images for GANs (Radford et al. 2015; Denton et al. 2015; Huang et al. 2017; Salimans et al. 2016; Odena et al. 2016; Arora et al. 2017; Berthelot et al. 2017; Zhang et al. 2018; Brock et al. 2018; Jolicoeur-Martineau 2018). Radford et al. (2015) first introduced convolutional layers to GANs architecture and proposed a network architecture called deep convolutional GANs (DCGANs). Denton et al. (2015) proposed a framework called Laplacian pyramid of GANs to improve the image quality of high-resolution images, where a Laplacian pyramid is constructed to generate high-resolution images from low-resolution images. A similar approach was proposed by Huang et al. (2017) who used a series of stacked GANs to generate images from abstract to specific. Salimans et al. (2016) proposed a technique called feature matching to achieve better convergence. The idea is to make the generated samples match the statistics of real data by minimizing the mean square error on an intermediate layer of the discriminator.

Training stability is another critical issue for GANs, and many studies have been proposed to address this problem by analyzing the objective functions of GANs (Arjovsky et al. 2017; Che et al. 2016; Metz et al. 2016; Nowozin et al. 2016; Qi 2017). Viewing the discriminator as an energy function, Zhao et al. (2016) used an autoencoder architecture to improve the stability of GANs learning. Dai et al. (2017) extended the energy-based GANs by adding some regularizations to make the discriminator non-degenerate. To make the generator and the discriminator more balanced, Metz et al. (2016) created an unrolled objective function to enhance the generator. Che et al. (2016) incorporated a reconstruction module and used the distance between real samples and reconstructed samples as a regularizer to obtain more stable gradients. Nowozin et al. (2016) pointed out that the objective of regular GAN (Goodfellow et al. 2014), which is related to Jensen-Shannon divergence, is a special case of divergence estimation and generalized it to arbitrary f-divergences (Nguyen et al. 2010). Arjovsky et al. (2017) extended this by analyzing the properties of four different divergences and concluded that Wasserstein distance is more stable than Jensen-Shannon divergence. Qi (2017) proposed the loss-sensitive GANs, whose loss function is based on the assumption that real samples should have smaller losses than fake samples. They also introduced the use of Lipschitz regularity to stabilize the learning process. Based on the above assumptions, they proved that loss-sensitive GANs have a non-vanishing gradient almost everywhere. Some other techniques to stabilize GANs learning include the second-order method (Mescheder et al. 2017a) and gradient penalty (Gulrajani et al. 2017; Kodali et al. 2017; Roth et al. 2017). Mescheder et al. (2017a) analyzed the convergence property of GANs from the perspective of the eigenvalues of the equilibrium and proposed a method to regularize the eigenvalues, which in turn leads to better training stability. Gulrajani et al. (2017) used gradient penalty to enforce the Lipschitz constraint in Wasserstein distance. They showed that this approach is more stable than the method used in Arjovsky et al. (2017). Unlike Gulrajani et al. (2017) who applied gradient penalty around the region between the real data and the fake data, Kodali et al. (2017) proposed to apply gradient penalty around the real data manifold only, which has the advantage that it is applicable to various GANs. Roth et al. (2017) derived a new gradient-based regularization by showing that the addition of noise to the discriminator yields a training set with gradient penalty.

2.2 Improving Quality for Generated Image with LSGANs

2.2.1 Least Squares Generative Adversarial Networks

As stated in Sect. 2.1.2, the original GANs (Goodfellow et al. 2014) use the sigmoid cross-entropy loss function for the discriminator and introduces the minimax loss and the non-saturating loss for the generator. However, both the minimax loss and the non-saturating loss will cause the problem of vanishing gradients for some fake samples that are far from real data, as shown in Fig. 2.4c. To remedy this problem, we propose the Least Squares GANs (LSGANs). Suppose we use the *a*–*b* coding scheme for the discriminator, where a and b are the labels for the fake data and the real data, respectively. Then the objective functions for LSGANs can be defined as follows:

$$\begin{aligned} \min_D V_{\text{LSGAN}}(D) =& \frac{1}{2}\mathbb{E}_{\boldsymbol{x}\sim p_{\text{data}}(\boldsymbol{x})}\big[(D(\boldsymbol{x})-b)^2\big] + \frac{1}{2}\mathbb{E}_{\boldsymbol{z}\sim p_{\boldsymbol{z}}(\boldsymbol{z})}\big[(D(G(\boldsymbol{z}))-a)^2\big], \\ \min_G V_{\text{LSGAN}}(G) =& \frac{1}{2}\mathbb{E}_{\boldsymbol{z}\sim p_{\boldsymbol{z}}(\boldsymbol{z})}\big[(D(G(\boldsymbol{z}))-c)^2\big], \end{aligned} \tag{2.4}$$

where c denotes the value that G wants D to believe for the fake data.

Unlike M-GANs and NS-GANs, which cause almost no gradient for some kinds of fake samples, LSGANs will penalize those samples even though they are correctly classified, as shown in Fig. 2.4d. When we update the generator, the parameters of the discriminator are fixed, that is, the decision boundary is fixed. As a result, the penalty will cause the generator to generate samples toward the decision boundary. On the other hand, the decision boundary should go across the manifold of real data for successful GANs learning; otherwise, the learning process will be saturated. Thus, moving the generated samples toward the decision boundary brings them closer to the manifold of the real data.

2.2.1.1 Theoretical Analysis

In the original GANs paper (Goodfellow et al. 2014), the authors showed that minimizing Equation (2.1) leads to a minimization of Jensen-Shannon divergence:

$$C(G) = \mathrm{KL}\left(p_{\text{data}} \middle\| \frac{p_{\text{data}} + p_g}{2}\right) + \mathrm{KL}\left(p_g \middle\| \frac{p_{\text{data}} + p_g}{2}\right) - \log(4). \tag{2.5}$$

Here we also explore the relation between LSGANs and f-divergence (Nowozin et al. 2016). Consider the following extension of Equation (2.4):

$$\begin{aligned}\min_D V_{\text{LSGAN}}(D) =& \frac{1}{2}\mathbb{E}_{\boldsymbol{x}\sim p_{\text{data}}(\boldsymbol{x})}\left[(D(\boldsymbol{x}) - b)^2\right] + \frac{1}{2}\mathbb{E}_{\boldsymbol{z}\sim p_{\boldsymbol{z}}(\boldsymbol{z})}\left[(D(G(\boldsymbol{z})) - a)^2\right],\\ \min_G V_{\text{LSGAN}}(G) =& \frac{1}{2}\mathbb{E}_{\boldsymbol{x}\sim p_{\text{data}}(\boldsymbol{x})}\left[(D(\boldsymbol{x}) - c)^2\right] + \frac{1}{2}\mathbb{E}_{\boldsymbol{z}\sim p_{\boldsymbol{z}}(\boldsymbol{z})}\left[(D(G(\boldsymbol{z})) - c)^2\right].\end{aligned} \tag{2.6}$$

Note that adding the term $\mathbb{E}_{\boldsymbol{x}\sim p_{\text{data}}(\boldsymbol{x})}[(D(\boldsymbol{x}) - c)^2]$ to $V_{\text{LSGAN}}(G)$ causes no change of the optimal values because this term does not contain parameters of G.

We first derive the optimal discriminator D for a fixed G.

Proposition 2.1 *For a fixed G, the optimal discriminator D is*

$$D^*(\boldsymbol{x}) = \frac{b p_{\text{data}}(\boldsymbol{x}) + a p_g(\boldsymbol{x})}{p_{\text{data}}(\boldsymbol{x}) + p_g(\boldsymbol{x})}. \tag{2.7}$$

Proof Given any generator G, we try to minimize $V(D)$ with respect to the discriminator D:

$$\begin{aligned}V(D) =& \frac{1}{2}\mathbb{E}_{\boldsymbol{x}\sim p_{\text{data}}}\left[(D(\boldsymbol{x}) - b)^2\right] + \frac{1}{2}\mathbb{E}_{\boldsymbol{z}\sim p_{\boldsymbol{z}}}\left[(D(G(\boldsymbol{z})) - a)^2\right]\\ =& \frac{1}{2}\mathbb{E}_{\boldsymbol{x}\sim p_{\text{data}}}\left[(D(\boldsymbol{x}) - b)^2\right] + \frac{1}{2}\mathbb{E}_{\boldsymbol{x}\sim p_g}\left[(D(\boldsymbol{x}) - a)^2\right]\\ =& \int_{\mathcal{X}} \frac{1}{2}\left(p_{\text{data}}(\boldsymbol{x})(D(\boldsymbol{x}) - b)^2 + p_g(\boldsymbol{x})(D(\boldsymbol{x}) - a)^2\right)\mathrm{d}x.\end{aligned} \tag{2.8}$$

Consider the internal function

$$\frac{1}{2}\left(p_{\text{data}}(\boldsymbol{x})(D(\boldsymbol{x}) - b)^2 + p_g(\boldsymbol{x})(D(\boldsymbol{x}) - a)^2\right). \tag{2.9}$$

It achieves the minimum at $\frac{b p_{\text{data}}(\boldsymbol{x}) + a p_g(\boldsymbol{x})}{p_{\text{data}}(\boldsymbol{x}) + p_g(\boldsymbol{x})}$ with respect to $D(\boldsymbol{x})$, concluding the proof.

In the following equations, we use p_{d} to denote p_{data} for simplicity.

Theorem 2.1 *Optimizing LSGANs yields minimizing the Pearson χ^2 divergence between $p_d + p_g$ and $2p_g$, if a, b, and c satisfy the conditions of $b - c = 1$ and $b - a = 2$ in Equation* (2.6).

Proof We can reformulate $V_{\mathrm{LSGAN}}(G)$ in Equation (2.6) by using Proposition 2.1:

$$
\begin{aligned}
2C(G) &= \mathbb{E}_{\boldsymbol{x}\sim p_{\mathrm{d}}}\left[(D^*(\boldsymbol{x}) - c)^2\right] + \mathbb{E}_{\boldsymbol{z}\sim p_z}\left[(D^*(G(\boldsymbol{z})) - c)^2\right] \\
&= \mathbb{E}_{\boldsymbol{x}\sim p_{\mathrm{d}}}\left[(D^*(\boldsymbol{x}) - c)^2\right] + \mathbb{E}_{\boldsymbol{x}\sim p_g}\left[(D^*(\boldsymbol{x}) - c)^2\right] \\
&= \mathbb{E}_{\boldsymbol{x}\sim p_{\mathrm{d}}}\left[\left(\frac{bp_{\mathrm{d}}(\boldsymbol{x}) + ap_g(\boldsymbol{x})}{p_{\mathrm{d}}(\boldsymbol{x}) + p_g(\boldsymbol{x})} - c\right)^2\right] \\
&\quad + \mathbb{E}_{\boldsymbol{x}\sim p_g}\left[\left(\frac{bp_{\mathrm{d}}(\boldsymbol{x}) + ap_g(\boldsymbol{x})}{p_{\mathrm{d}}(\boldsymbol{x}) + p_g(\boldsymbol{x})} - c\right)^2\right] \\
&= \int_{\mathcal{X}} p_{\mathrm{d}}(\boldsymbol{x})\left(\frac{(b-c)p_{\mathrm{d}}(\boldsymbol{x}) + (a-c)p_g(\boldsymbol{x})}{p_{\mathrm{d}}(\boldsymbol{x}) + p_g(\boldsymbol{x})}\right)^2 \mathrm{d}\boldsymbol{x} \\
&\quad + \int_{\mathcal{X}} p_g(\boldsymbol{x})\left(\frac{(b-c)p_{\mathrm{d}}(\boldsymbol{x}) + (a-c)p_g(\boldsymbol{x})}{p_{\mathrm{d}}(\boldsymbol{x}) + p_g(\boldsymbol{x})}\right)^2 \mathrm{d}\boldsymbol{x} \\
&= \int_{\mathcal{X}} \frac{\left((b-c)p_{\mathrm{d}}(\boldsymbol{x}) + (a-c)p_g(\boldsymbol{x})\right)^2}{p_{\mathrm{d}}(\boldsymbol{x}) + p_g(\boldsymbol{x})} \mathrm{d}\boldsymbol{x} \\
&= \int_{\mathcal{X}} \frac{\left((b-c)(p_{\mathrm{d}}(\boldsymbol{x}) + p_g(\boldsymbol{x})) - (b-a)p_g(\boldsymbol{x})\right)^2}{p_{\mathrm{d}}(\boldsymbol{x}) + p_g(\boldsymbol{x})} \mathrm{d}\boldsymbol{x}.
\end{aligned}
\tag{2.10}
$$

If we set $b - c = 1$ and $b - a = 2$, then

$$
\begin{aligned}
2C(G) &= \int_{\mathcal{X}} \frac{\left(2p_g(\boldsymbol{x}) - (p_{\mathrm{d}}(\boldsymbol{x}) + p_g(\boldsymbol{x}))\right)^2}{p_{\mathrm{d}}(\boldsymbol{x}) + p_g(\boldsymbol{x})} \mathrm{d}\boldsymbol{x} \\
&= \chi^2_{\mathrm{Pearson}}(p_{\mathrm{d}} + p_g \| 2p_g),
\end{aligned}
\tag{2.11}
$$

where $\chi^2_{\mathrm{Pearson}}$ is the Pearson χ^2 divergence. Thus, minimizing Equation (2.6) leads to a minimization of Pearson χ^2 divergence between $p_{\mathrm{d}} + p_g$ and $2p_g$, if a, b, and c satisfy the conditions of $b - c = 1$ and $b - a = 2$.

2.2.1.2 Parameter Selection

One method to determine the values of a, b, and c in Equation (2.4) is to satisfy the conditions of $b - c = 1$ and $b - a = 2$, such that minimizing Equation (2.4) leads to a minimization of Pearson χ^2 divergence between $p_{\rm d} + p_g$ and $2p_g$. For example, by setting $a = -1$, $b = 1$, and $c = 0$, we get the following objective functions:

$$\begin{aligned}\min_D V_{\rm LSGAN}(D) =& \frac{1}{2}\mathbb{E}_{\boldsymbol{x}\sim p_{\rm data}(\boldsymbol{x})}\big[(D(\boldsymbol{x})-1)^2\big] + \frac{1}{2}\mathbb{E}_{\boldsymbol{z}\sim p_{\boldsymbol{z}}(\boldsymbol{z})}\big[(D(G(\boldsymbol{z}))+1)^2\big],\\ \min_G V_{\rm LSGAN}(G) =& \frac{1}{2}\mathbb{E}_{\boldsymbol{z}\sim p_{\boldsymbol{z}}(\boldsymbol{z})}\big[(D(G(\boldsymbol{z})))^2\big].\end{aligned} \tag{2.12}$$

Another method is to make G generate samples as real as possible by setting $c = b$, corresponding to the traditional method of using least squares for classification. For example, by using the 0–1 binary coding scheme, we obtain the following objective functions:

$$\begin{aligned}\min_D V_{\rm LSGAN}(D) =& \frac{1}{2}\mathbb{E}_{\boldsymbol{x}\sim p_{\rm data}(\boldsymbol{x})}\big[(D(\boldsymbol{x})-1)^2\big] + \frac{1}{2}\mathbb{E}_{\boldsymbol{z}\sim p_{\boldsymbol{z}}(\boldsymbol{z})}\big[(D(G(\boldsymbol{z})))^2\big],\\ \min_G V_{\rm LSGAN}(G) =& \frac{1}{2}\mathbb{E}_{\boldsymbol{z}\sim p_{\boldsymbol{z}}(\boldsymbol{z})}\big[(D(G(\boldsymbol{z}))-1)^2\big].\end{aligned} \tag{2.13}$$

In practice, we find that Equation (2.12) shows better FID results and a faster convergence speed than Equation (2.13), as demonstrated by experiments.

2.2.2 Experiments

2.2.2.1 Implementation Details

The implementation of our proposed models is based on a public implementation of DCGANs[1] using TensorFlow (Abadi et al. 2016). The learning rate is set to 0.0002, except for LSUN-scenes, whose learning rate is set to 0.001. The mini-batch size is set to 64, and the variables are initialized from a Gaussian distribution with a mean of zero and a standard deviation of 0.02. Following DCGANs, β_1 for the Adam optimizer is set to 0.5. The pixel values of all the images are scaled to $[-1,1]$, because we use the Tanh in the generator to produce images.

2.2.2.2 Qualitative Evaluation

Scene Generation

We train LSGANs and NS-GANs with the same network architecture as on the LSUN-bedroom dataset. The network architecture is presented in Table 2.1. All

[1] https://github.com/carpedm20/DCGAN-tensorflow

Table 2.1 Network architecture for scene generation, where CONV denotes the convolutional layer, TCONV denotes the transposed convolutional layer, FC denotes the fully-connected layer, BN denotes batch normalization, LReLU denotes the Leaky-ReLU, and (K3,S2,O256) denotes a layer with 3 × 3 kernel, stride 2, and 256 output filters

Generator	Discriminator
Input z	Input 112 × 112 × 3
FC(O12544), BN, ReLU	CONV(K5,S2,O64), LReLU
TCONV(K3,S2,O256), BN, ReLU	CONV(K5,S2,O128), BN, LReLU
TCONV(K3,S1,O256), BN, ReLU	CONV(K5,S2,O256), BN, LReLU
TCONV(K3,S2,O256), BN, ReLU	CONV(K5,S2,O512), BN, LReLU
TCONV(K3,S1,O256), BN, ReLU	FC(O1)
TCONV(K3,S2,O128), BN, ReLU	Loss
TCONV(K3,S2,O64), BN, ReLU	
TCONV(K3,S1,O3), Tanh	

images are resized to a resolution of 112 × 112. The images generated by the two models are presented in Fig. 2.5. Compared with the images generated by NS-GANs, the texture details (e.g., the textures of beds) of the images generated by LSGANs are more exquisite and look sharper. We also train LSGANs on four other scene datasets, including the church, the dining room, the kitchen, and the conference room. The generated results are shown in Fig. 2.6.

Cat Generation

We further evaluate LSGANs on a cat dataset (Zhang et al. 2008). We first use the preprocess methods in a public project[2] to obtain cat head images with a resolution larger than 128 × 128 and then resize all the images to a resolution of 128 × 128. The network architecture used in this task is presented in Table 2.2. We use the following evaluation protocol to compare the performance between LSGANs and NS-GANs. First, we train LSGANs and NS-GANs using the same architecture on the cat dataset. During training, we save a checkpoint of the model and a batch of the generated images every 1000 iterations. Second, we select the best models of LSGANs and NS-GANs by checking the quality of saved images in every 1000 iterations. Finally, we use the selected best models to randomly generate cat images and compare the quality of the generated images. Figure 2.7 shows the generated cat images of LSGANs and NS-GANs. We observe that LSGANs generate cats with sharper and more exquisite hair than the ones generated by NS-GANs. Figure 2.7c and d shows the details of the cat hair by zooming in the generated images. We

[2]https://github.com/AlexiaJM/Deep-learning-with-cats

Fig. 2.5 Images generated from LSUN-bedroom. (**a**) Images (64 × 64) generated by NS-GANs (reported in Radford et al. 2015). (**b**) Images (112 × 112) generated by NS-GANs. (**c**) Images (112 × 112) generated by LSGANs

Fig. 2.6 Images generated from different scene datasets. (**a**) Church outdoor. (**b**) Dining room. (**c**) Kitchen. (**d**) Conference room

Table 2.2 Network architecture for cat generation. The symbols are defined in Table 2.1

Generator	Discriminator
Input z	Input 128 × 128 × 3
FC(O32768), BN, ReLU	CONV(K5,S2,O64), LReLU
TCONV(K3,S2,O256), BN, ReLU	CONV(K5,S2,O128), BN, LReLU
TCONV(K3,S2,O128), BN, ReLU	CONV(K5,S2,O256), BN, LReLU
TCONV(K3,S2,O64), BN, ReLU	CONV(K5,S2,O512), BN, LReLU
TCONV(K3,S2,O3), Tanh	FC(O1)
	Loss

observe that the cat hair generated by NS-GANs contains more artificial noise. By checking more generated samples using the above-saved models, we also observe that the overall quality of the images generated by LSGANs is better than that of NS-GANs.

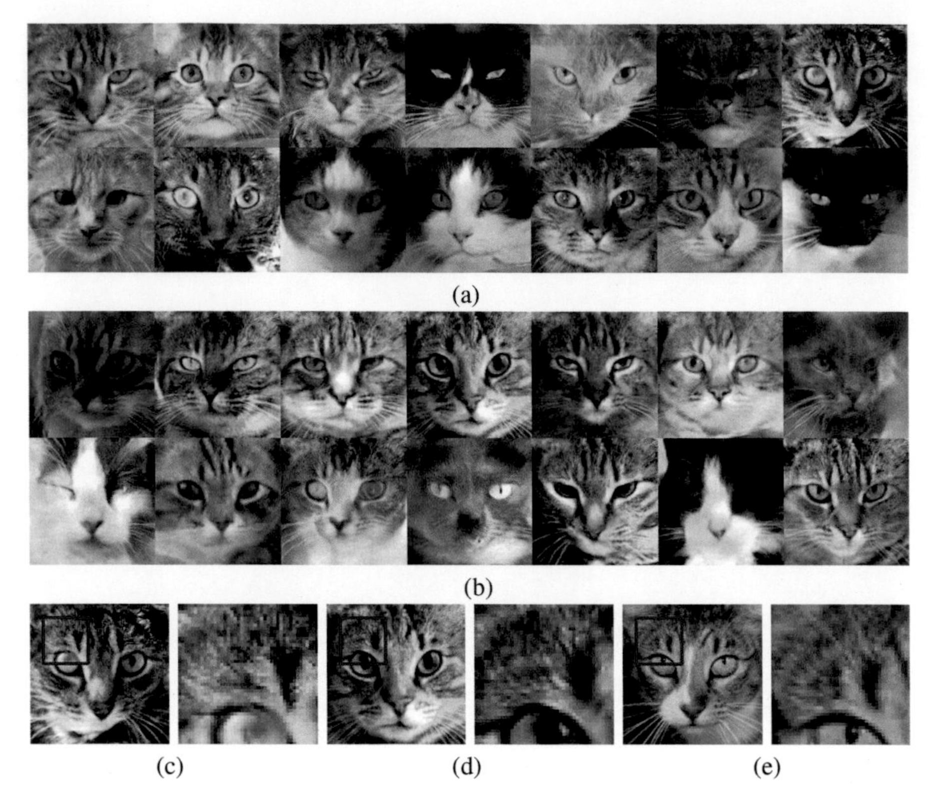

Fig. 2.7 Generated images on cat datasets. (**c**, **d**, **e**): Comparison by zooming in on the details of the images. LSGANs generate cats with sharper and more exquisite hair and faces than the ones generated by NS-GANs. (**a**) Generated cats (128 × 128) by NS-GANs. (**b**) Generated cats (128 × 128) by LSGANs. (**c**) NS-GANs. (**d**) LSGANs. (**e**) Real Sample

Walking in the Latent Space

We also present the interpolation results in Fig. 2.8. The result of walking in the latent space is a sign of whether a model is simply memorizing the training dataset. We first randomly sample two points of the noise vector z and then interpolate the vector values between the two sampled points. The images in Fig. 2.8 show smooth transitions, which indicates that LSGANs have learned semantic representations in the latent space.

2.2.2.3 Quantitative Evaluation

For the quantitative evaluation of LSGANs, we adopt Fréchet Inception Distance (FID) (Heusel et al. 2017) as the evaluation metric. FID measures the distance between the generated images and the real images by approximating the feature

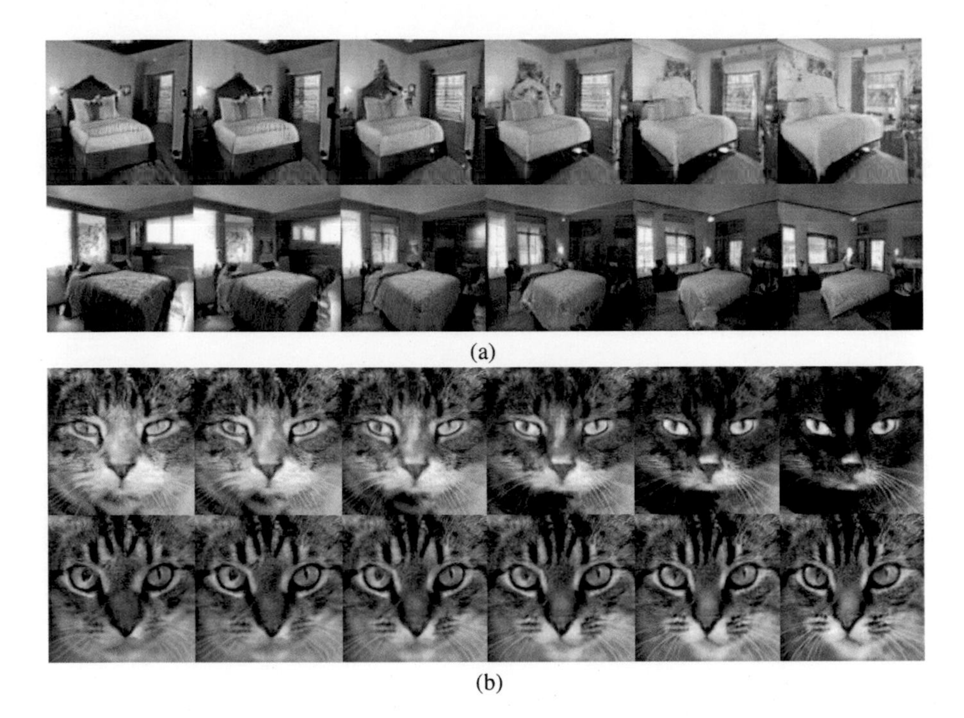

(a)

(b)

Fig. 2.8 Interpolation result by LSGANs. The generated images show smooth transitions, which indicates that LSGANs have learned semantic representations in the latent space. (**a**) Interpolation on the LSUN-bedroom dataset. (**b**) Interpolation on the cat dataset

space of the inception model as a multidimensional Gaussian distribution, which has been proved to be more consistent with human judgment than inception score (Salimans et al. 2016). Smaller FID values mean shorter distances between the generated and real images. We also conduct a human subjective study on the LSUN-bedroom dataset.

Fréchet Inception Distance

For FID, we evaluate the performances of LSGANs, NS-GANs, and WGANs-GP on several datasets, including LSUN-bedroom, the cat dataset, ImageNet, and CIFAR-10. We also compare the performances of Equation (2.12) (denoted as LSGANs$_{(-110)}$) and Equation (2.13) (denoted as LSGANs$_{(011)}$). For a fair comparison, all models are trained with the same architecture proposed in DCGANs (Radford et al. 2015) (i.e., four convolutional layers for both the discriminator and the generator), and the dimension of the noise input is set to 100. For WGANs-GP, we adopt the official implementation for evaluation. The resolutions for LSUN, Cat, ImageNet, and CIFAR-10 are 64×64, 128×128, 64×64, and 32×32, respectively. We randomly generate 50,000 images every 4000 iterations for each model and then compute the FID. The results are shown in Table 2.3, and we have the following four observations. First, LSGANs$_{(-110)}$ outperform NS-GANs for all four

Table 2.3 FID results of NS-GANs, WGANs-GP, and LSGANs on four datasets. LSGANs$_{(-110)}$ and LSGANs$_{(011)}$ refer to Equations (2.12) and (2.13), respectively

Method	LSUN	Cat	ImageNet	CIFAR10
NS-GANs	28.04	15.81	74.15	35.25
WGANs-GP	22.77	29.03	**62.05**	40.83
LSGANs$_{(011)}$	27.21	15.46	72.54	36.46
LSGANs$_{(-110)}$	**21.55**	**14.28**	68.95	**35.19**

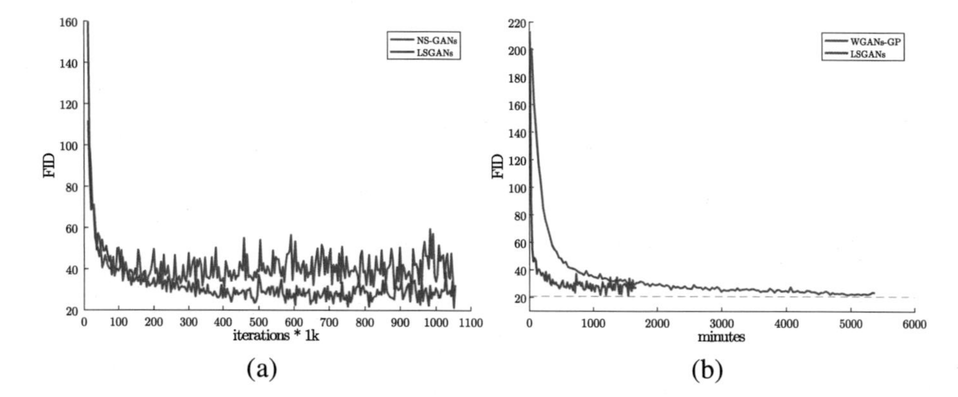

Fig. 2.9 (**a**): Comparison of FID on LSUN between NS-GANs and LSGANs during the learning process, which is aligned with iterations. (**b**): Comparison of FID on LSUN between WGANs-GP and LSGANs during the learning process, which is aligned with wall-clock time

datasets. Second, compared with WGANs-GP, LSGANs$_{(-110)}$ perform better for three datasets, especially for the cat dataset. Third, LSGANs$_{(-110)}$ perform better than LSGANs$_{(011)}$ for all four datasets. Fourth, the performance of LSGANs$_{(011)}$ is comparable to that of NS-GANs.

We also show the FID plot of the learning process in Fig. 2.9, where LSGANs refer to LSGANs$_{(-110)}$. Following (Heusel et al. 2017), the plots of NS-GANs and LSGANs are aligned by iterations, and the plots of WGANs-GP and LSGANs are aligned by wall-clock time. As Fig. 2.9a shows, NS-GANs and LSGANs show similar FID at the first 25,000 iterations, but LSGANs can decrease FID after 25,000 iterations, eventually achieving better performance. Figure 2.9b shows that WGANs-GP and LSGANs achieve similar optimal FID eventually, but LSGANs spend much less time (1100 minutes) than WGANs-GP (4600 minutes) to reach a relatively optimal FID around 22. This is because WGANs-GP need multiple updates for the discriminator and need additional computational time for the gradient penalty.

Human Subjective Study

To further evaluate the performance of LSGANs, we conduct a human subjective study using the generated bedroom images (112 × 112) from NS-GANs and

LSGANs with the same network architecture. We randomly construct image pairs, where one image is from NS-GANs, and the other is from LSGANs. We asked Amazon Mechanical Turk annotators to judge which image looks more realistic. With 4000 votes, NS-GANs received get 43.6% votes, and LSGANs received 56.4% votes, that is, an overall 12.8% increase of votes over NS-GANs.

2.2.2.4 Comparison of Two Parameter Schemes

As stated in Sect. 2.2.2.3, LSGANs$_{(-110)}$ perform better than LSGANs$_{(011)}$ for the FID-based experiment. In this experiment, we show another comparison between the two parameter schemes. We train LSGANs$_{(-110)}$ and LSGANs$_{(011)}$ on SVHN (Netzer et al. 2011) dataset using the same network architecture. Figure 2.10 shows the dynamic results of the two schemes. We can observe that LSGANs$_{(-110)}$ show a faster convergence speed than LSGANs$_{(011)}$. We also evaluate the two schemes on the LSUN-bedroom and cat datasets, and similar results are observed.

Discussion
Based on the FID results and the convergence speed of LSGANs$_{(011)}$ and LSGANs$_{(-110)}$, we can conclude that LSGANs$_{(-110)}$ perform better than

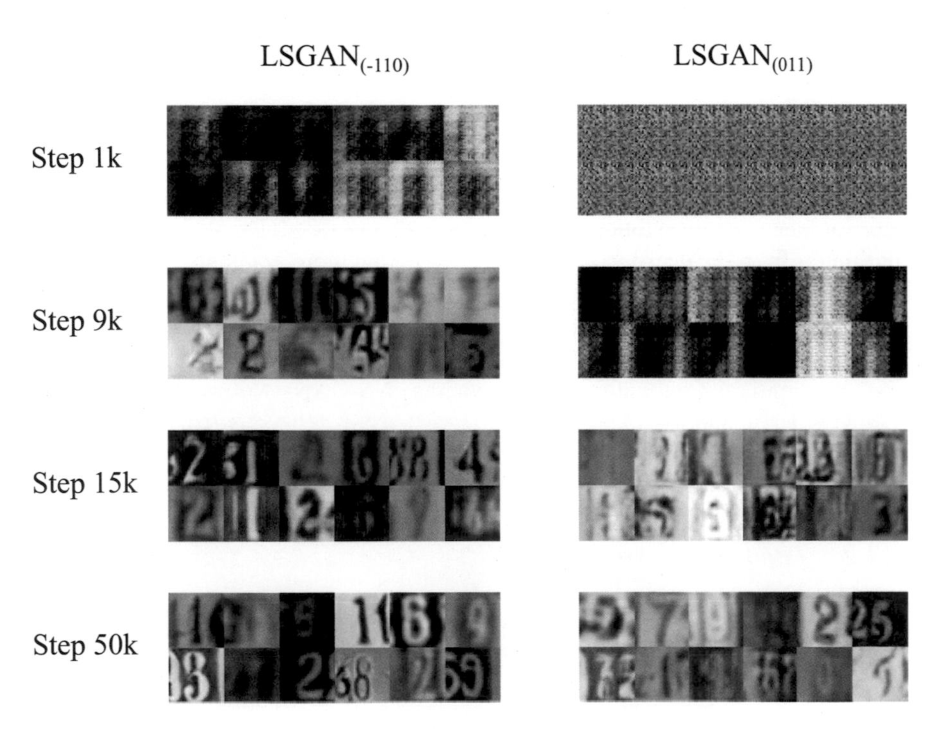

Fig. 2.10 Dynamic results of the two parameter schemes on the SVHN dataset. The first row corresponds to Equation (2.12), and the second row corresponds to Equation (2.13)

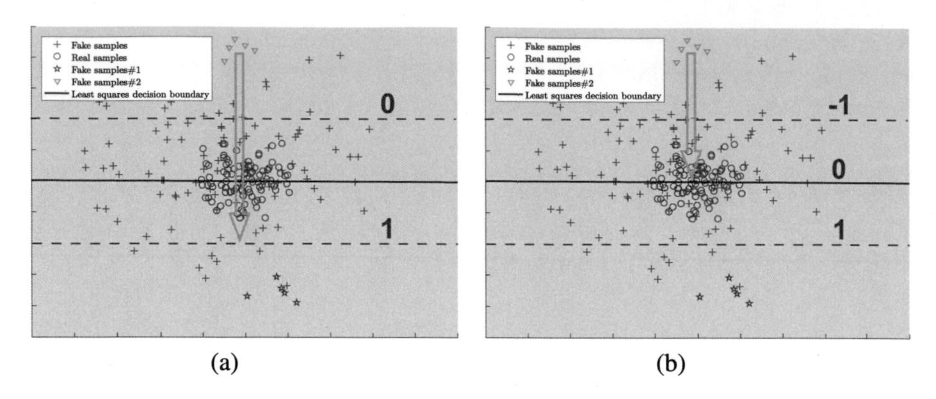

Fig. 2.11 Explanation of two parameter schemes. (**a**) LSGANs$_{(011)}$. (**b**) LSGANs$_{(-110)}$

LSGANs$_{(011)}$. An intuitive explanation is presented in Fig. 2.11. For LSGANs$_{(-110)}$, the fake samples will be pulled toward boundary "0," which goes across the manifold of the real data. However, for LSGANs$_{(011)}$, the fake samples will be pulled toward boundary "1." Thus, LSGANs$_{(-110)}$ can generate samples that are closer to real samples than LSGANs$_{(011)}$.

2.3 Improving Training Stability: Theoretical Analysis

2.3.1 Approach

As stated in Sect. 2.2.1, LSGANs penalize samples based on their distances to the decision boundary even though they are correctly classified. Penalizing the samples that lie far from the decision boundary can generate more gradients when updating the generator, which in turn relieves the problem of vanishing gradients. This allows LSGANs to perform more stable during the learning process. This benefit can also be derived from another perspective: as shown in Fig. 2.12, the least squares loss function is flat only at one point, whereas NS-GANs will saturate when x is relatively large, and M-GANs will saturate when x is relatively small. We provide further theoretical analysis about the stability of LSGANs in the following section.

2.3.2 Theoretical Analysis

As stated in Sect. 2.2.1.1, the original GANs have been proven to optimize the JS divergence:

$$C(G) = \mathrm{KL}\left(p_{\text{data}} \left\| \frac{p_{\text{data}} + p_g}{2}\right.\right) + \mathrm{KL}\left(p_g \left\| \frac{p_{\text{data}} + p_g}{2}\right.\right) - \log(4). \tag{2.14}$$

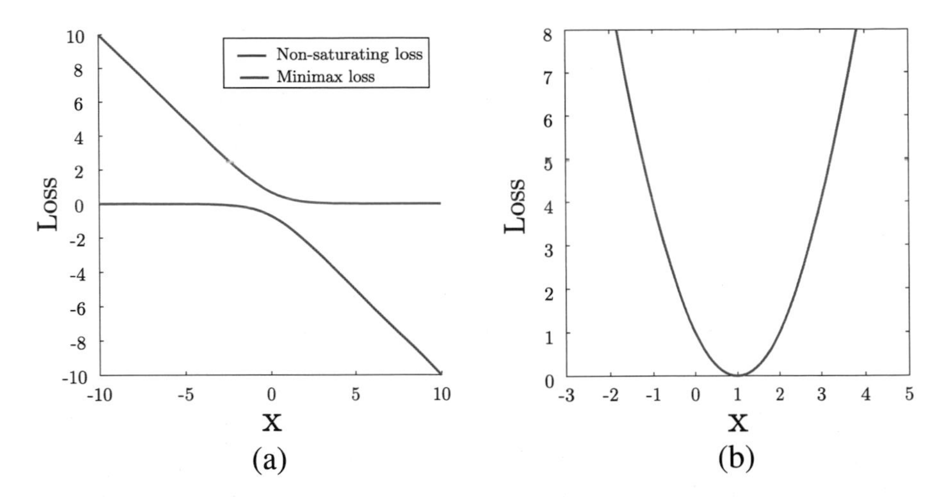

Fig. 2.12 (**a**): Non-saturating loss and the minimax loss. (**b**): Least squares loss

Furthermore, (Huszar 2015) pointed out that Equation (2.14) can be viewed as an interpolation between $\mathrm{KL}(p_g \| p_d)$ and $\mathrm{KL}(p_d \| p_g)$:

$$\mathrm{JS}_\pi(p_d \| p_g) = (1-\pi)\mathrm{KL}(p_d \| \pi p_d + (1-\pi)p_g) + \pi \mathrm{KL}(p_g \| \pi p_d + (1-\pi)p_g), \tag{2.15}$$

where Equation (2.14) corresponds to $\pi = 0.5$. They also found that optimizing Equation (2.14) tends to perform similarly to $\mathrm{KL}(p_g \| p_d)$. $\mathrm{KL}(p_g \| p_d)$ is widely used in variational inference due to the convenient evidence lower bound (Bishop 2006). However, optimizing $\mathrm{KL}(p_g \| p_d)$ has the problem of mode-seeking behavior or under-dispersed approximations (Bishop 2006; Dieng et al. 2017; Huszar 2015). This problem also appears in GANs learning, which is known as the mode collapse problem. The definition of $\mathrm{KL}(p_g \| p_d)$ is given below:

$$\mathrm{KL}(p_g \| p_d) = -\int_{\mathcal{X}} p_g(\boldsymbol{x}) \ln \left(\frac{p_d(\boldsymbol{x})}{p_g(\boldsymbol{x})} \right) \mathrm{d}\boldsymbol{x}. \tag{2.16}$$

The mode-seeking behavior of $\mathrm{KL}(p_g \| p_d)$ can be understood by noting that p_g will approach zero where p_d is near zero, because $\mathrm{KL}(p_g \| p_d)$ will be infinite if $p_d = 0$ and $p_g > 0$. This is called the zero-forcing property (Bishop 2006).

In Sect. 2.2.1.1, we have proved that minimizing Equation 2.12 minimizes the Pearson χ^2 divergence between $p_\mathrm{d} + p_g$ and $2p_g$:

$$\begin{aligned} 2C(G) &= \int_{\mathcal{X}} \frac{\left(2p_g(\boldsymbol{x}) - (p_\mathrm{d}(\boldsymbol{x}) + p_g(\boldsymbol{x}))\right)^2}{p_\mathrm{d}(\boldsymbol{x}) + p_g(\boldsymbol{x})} \mathrm{d}\boldsymbol{x} \\ &= \chi^2_{\mathrm{Pearson}}(p_\mathrm{d} + p_g \| 2p_g). \end{aligned} \tag{2.17}$$

Recently, χ^2 divergence has drawn researchers' attention in variational inference because χ^2 divergence can produce over-dispersed approximations (Dieng et al. 2017). For the objective function in Equation (2.11), it will become infinite if $p_d + p_g = 0$ and $p_g - p_d > 0$, which will not happen because $p_g \geq 0$ and $p_d \geq 0$. Thus, $\chi^2_{\text{Pearson}}(p_d + p_g \| 2p_g)$ has no zero-forcing property. This makes LSGANs less mode-seeking and alleviates the mode collapse problem.

2.3.3 Experiments

In this section, we evaluate the stability of our proposed LSGANs and compare them with two baselines including NS-GANs and WGANs-GP. Gradient penalty has been proven to be effective for improving the stability of GANs training (Kodali et al. 2017; Gulrajani et al. 2017), but it also has some inevitable disadvantages such as additional computational cost and memory cost. Thus, we evaluate the stability of LSGANs in two groups. One is used to compare with the model without gradient penalty (i.e., NS-GANs), and the other is used to compare with the model with gradient penalty (i.e., WGANs-GP).

2.3.3.1 Evaluation Without Gradient Penalty

We first compare LSGANs with NS-GANs, neither of which has a gradient penalty. Three comparison experiments are conducted: (1) learning on a Gaussian mixture distribution, (2) learning with difficult architectures, and (3) learning on datasets with small variability.

2.3.3.2 Gaussian Mixture Distribution

Learning on a Gaussian mixture distribution to evaluate stability was proposed by Metz et al. (2016). If the model suffers from the mode collapse problem, it will generate samples only around one or two modes. We train NS-GANs and LSGANs with the same network architecture on a 2D mixture of eight Gaussian mixture distribution, where both the generator and the discriminator contain three fully connected layers. Figure 2.13 shows the dynamic results of Gaussian kernel density estimation. We can see that NS-GANs suffer from mode collapse starting at 15000 iterations. They only generate samples around a single valid mode of the data distribution. But LSGANs can learn the Gaussian mixture distribution successfully. We also try different architectures (four or five fully connected layers) and different values of the hyper-parameters (the learning rate and the dimension of the noise vector). The results also show that NS-GANs tend to generate samples around one or two modes, whereas LSGANs are less prone to this problem.

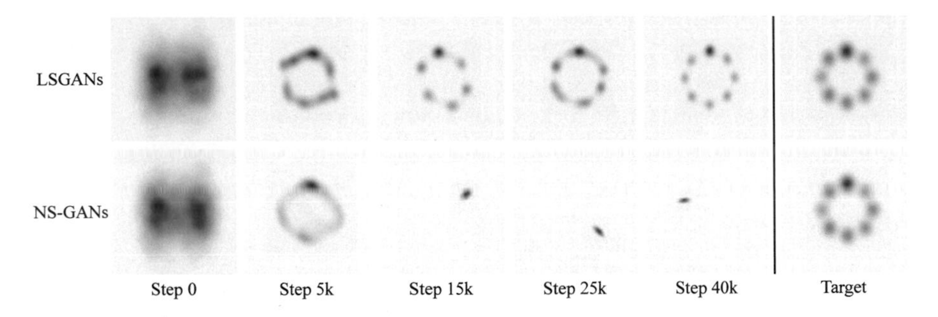

Fig. 2.13 Dynamic results of Gaussian kernel estimation for NS-GANs and LSGANs. The final column shows the distribution of real data

Table 2.4 Experiments on Gaussian mixture distribution. We run 100 times for each model and record how many times that a model ever generates samples around one or two modes during the training process

Method	Number of generating samples around one or two modes
NS-GANs	99/100
LSGANs (ours)	5/100

To further verify the robustness of the above observation, we run each model 100 times and record how many times a model suffers from the mode collapse problem. For each experiment, we save the density estimation every 5000 iterations and observe whether a model generates samples only around one or two modes in each saved estimation. The results show that NS-GANs appear to generate one or two modes 99 times out of 100, whereas LSGANs do so only 5 times, as shown in Table 2.4.

2.3.3.3 Difficult Architectures

Another experiment is to train GANs with difficult architectures, which was proposed by Arjovsky et al. (2017). The model will generate very similar images if it suffers from the mode collapse problem. The network architecture used in this task is similar to the one in Table 2.2 except for the image resolution. Based on this network architecture, two architectures are designed to compare stability. The first is to exclude the batch normalization in the generator (BN_G for short), and the second is to exclude the batch normalization in both the generator and discriminator (BN_{GD} for short). As pointed out by Arjovsky et al. (2017), the selection of optimizer is critical to the model performance. Thus we evaluate the two architectures with two optimizers, Adam (Kingma and Ba 2014) and RMSProp (Tieleman and Hinton 2012). In summary, we have the following four training settings: (1) BN_G with Adam, (2) BN_G with RMSProp, (3) BN_{GD} with

Adam, and (4) BN_{GD} with RMSProp. We train the above models on the LSUN-bedroom dataset using NS-GANs and LSGANs separately. The results are shown in Fig. 2.14, and we make three major observations. First, for BN_G with Adam, there is a chance for LSGANs to generate relatively good-quality images. We test 10 times, and 5 of those succeeded in generating relatively good-quality images. For NS-GANs, however, we never observe successful learning due to the severe degree of mode collapse. Second, for BN_{GD} with RMSProp, as Fig. 2.14 shows, LSGANs generate higher-quality images than NS-GANs, which have a slight degree of mode collapse. Third, LSGANs and NS-GANs have similar performance for BN_G with RMSProp and BN_{GD} with Adam. Specifically, for BN_G with RMSProp, both LSGANs and NS-GANs can generate relatively good images. For BN_{GD} with Adam, both have a slight degree of mode collapse.

2.3.3.4 Datasets with Small Variability

The use of difficult architectures is an effective way to evaluate the stability of GANs (Arjovsky et al. 2017). However, in practice, it is natural to select a stable architecture for a given task. The difficulty of a practical task is the task itself. Inspired by this motivation, we propose to use difficult datasets but stable architectures to evaluate the stability of GANs. We find that datasets with small variability are difficult for GANs to learn, because the discriminator can distinguish the real samples very easily for the datasets with low variability. Specifically, we construct the datasets by rendering 28×28 digits using the Times New Roman font. Two datasets are created:[3] (1) one with a random horizontal shift and (2) the other one with a random horizontal shift and random rotation from 0 to 10 degrees. Each category contains 1000 samples for both datasets. Note that the second dataset has greater variability than the first one. Examples of the two synthetic datasets are shown in the first column of Fig. 2.15. We adopt a stable architecture for digit generation, following the suggestions in Radford et al. (2015), where the discriminator is similar to LeNet and the generator contains three transposed convolutional layers. The detail of the network architecture is presented in Table 2.5. We train NS-GANs and LSGANs on the above two datasets, and the generated images are shown in Fig. 2.15, along with the results on MNIST. We make two major observations. First, NS-GANs succeed in learning on MNIST but fail on the two synthetic digit datasets, whereas LSGANs succeed in learning on all the three datasets. Second, LSGANs generate higher-quality images on the second dataset than on the first, which implies that increasing the variability of the dataset can improve the generated image quality and relieve the mode collapse problem. Based on this observation, applying data augmentation such as shifting, cropping, and rotation is an effective way to improve GANs learning.

[3] Available at https://github.com/xudonmao/improved_LSGAN

Fig. 2.14 Comparison experiments between NS-GANs and LSGANs by excluding batch normalization (BN). (**a**) No BN in G using Adam. (**b**) No BN in either G or D using RMSProp. (**c**) No BN in G using RMSProp. (**d**) No BN in either G or D using Adam

2.3.3.5 Evaluation with Gradient Penalty

Gradient penalty has been proven to be effective in improving the stability of GANs training (Kodali et al. 2017; Gulrajani et al. 2017). To compare with WGANs-GP, which is the state-of-the-art model in terms of stability, we adopt the gradient

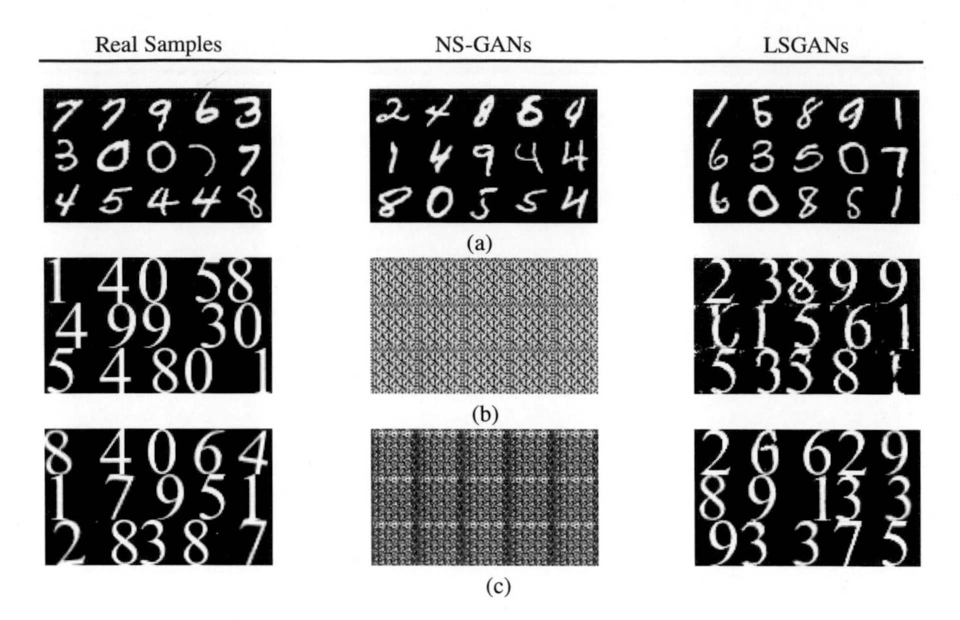

Fig. 2.15 Evaluation of datasets with small variability. All tasks are conducted using the same network architecture as shown in Table 2.5. DCGANs succeed in learning on MNIST but fail on the two synthetic digit datasets with small variability, whereas LSGANs succeed in learning on all the three datasets. (**a**) Training on MNIST. (**b**) Training on a synthetic digit dataset with a random horizontal shift. (**c**) Training on a synthetic digit dataset with random horizontal shift and rotation

Table 2.5 Network architecture for stability evaluation on datasets with small variability. The symbols are defined in Table 2.1

Generator	Discriminator
Input z	Input 28 × 28 × 1
FC(O8192), BN, ReLU	CONV(K5,S2,O20), LReLU
TCONV(K3,S2,O256), BN, ReLU	CONV(K5,S2,O50), BN, LReLU
TCONV(K3,S2,O128), BN, ReLU	FC(O500), BN, LReLU
TCONV(K3,S2,O1), Tanh	FC(O1)
	Loss

penalty proposed by Kodali et al. (2017) for LSGANs and set the hyper-parameters c and λ to 30 and 150, respectively. For this experiment, our implementation is based on the official implementation of WGANs-GP. We follow the evaluation method in WGANs-GP to train with six difficult architectures including (1) no normalization and a constant number of filters in the generator, (2) a 4-layer 512-dimension ReLU MLP generator, (3) no normalization in either the generator or discriminator, (4) gated multiplicative nonlinearities in both the generator and discriminator, (5) tanh nonlinearities in both the generator and discriminator, and (6) 101-layer ResNet for both the generator and discriminator. The results are presented in Fig. 2.16, where

Fig. 2.16 Comparison experiments between WGANs-GP and LSGANs-GP using difficult architectures, where the images generated by WGANs-GP are duplicated from Gulrajani et al. (2017). LSGANs-GP succeed for all architectures. (**a**) G: No BN and a constant number of filters, D: DCGAN. (**b**) G: 4-layer 512-dim ReLU MLP, D: DCGAN. (**c**) No normalization in either G or D. (**d**) Gated multiplicative nonlinearities everywhere in G and D. (**e**) Tanh nonlinearities everywhere in G and D. (**f**) 101-layer ResNet G and D

the images generated by WGANs-GP are duplicated from Gulrajani et al. (2017). We make the following two major observations. First, like WGANs-GP, LSGANs-GP also succeed in training for each architecture, including the 101-layer ResNet. Second, LSGANs-GP with 101-layer ResNet generate higher-quality images than the other five architectures.

2.3.4 Discussion

Based on the above experiments, we have the following suggestions in practice. First, we suggest the use of LSGANs$_{(-110)}$ without gradient penalty if it works, because the use of gradient penalty will introduce an additional computational cost and memory cost. Second, we observe that the quality of images generated by LSGANs may shift between good and bad during the training process, which is also indicated in Fig. 2.9. Thus, a record should be kept of generated images at every thousand or hundred iterations, and the model should be selected manually by checking the image quality. Third, if LSGANs without gradient penalty fail, we suggest the use of LSGANs-GP with the hyper-parameters according to the suggestions in the literature (Kodali et al. 2017). In our experiments, we find that the hyper-parameter setting ($c = 30$ and $\lambda = 150$) works for all tasks.

2.4 Multi-domain Image Generation with RCGANs

Multi-domain image generation is an important extension of image generation. The target of multi-domain image generation is to generate aligned image pairs, given two or more domain images. For example, in Fig. 2.17, the training data contain two domain images: male and female. After training, the model should be able to generate aligned image pairs that share a very similar appearance. Multi-domain image generation has many promising applications such as improving the generated image quality (Dosovitskiy et al. 2015; Wang and Gupta 2016), image-to-image translation (Perarnau et al. 2016; Wang et al. 2017a), and unsupervised domain adaptation (Liu and Tuzel 2016).

Several early approaches (Dosovitskiy et al. 2015; Wang and Gupta 2016) have been proposed, but they are all in the supervised setting, which means that they require the information of paired samples to be available. In practice, however, building paired training datasets can be very expensive and may not always be feasible.

Recently, CoGANs (Liu and Tuzel 2016) have achieved great success in multi-domain image generation. In particular, CoGANs model the problem as learning a joint distribution over multi-domain images by coupling multiple GANs. Unlike previous methods that require paired training data, CoGANs can learn the joint

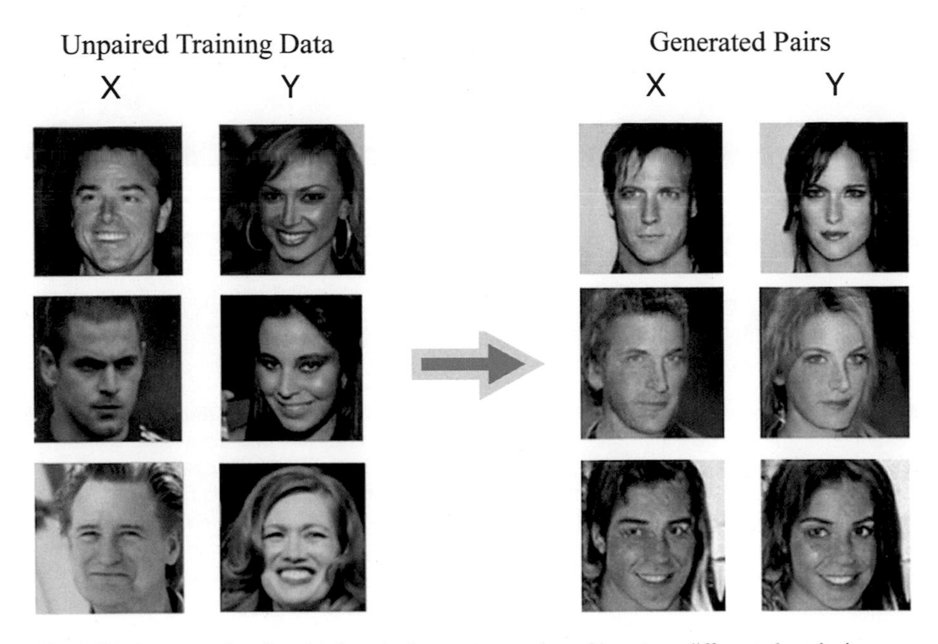

Fig. 2.17 An example of multi-domain image generation. Given two different domain images, such as male and female, the model should be able to generate aligned image pairs that share very similar appearances

distribution without any paired samples. However, it falls short for some difficult tasks such as the generation of edges and photos, as demonstrated by experiments.

We propose a new framework called Regularized Conditional GANs (RCGANs). Like CoGANs, RCGANs can perform multi-domain image generation in the absence of paired samples. RCGANs are based on the conditional GANs (Mirza and Osindero 2014) and try to learn a conditional distribution over multi-domain images, where the domain-specific semantics are encoded in the conditioned domain variables and the common semantics are encoded in the shared latent variables.

As pointed out by Liu and Tuzel (2016), the direct use of conditional GANs will fail to learn the corresponding semantics. To overcome this problem, we propose two regularizers to guide the model to encode the common semantics in the shared latent variables, which in turn makes the model generate corresponding images. As shown in Fig. 2.18a and b, one regularizer is used in the first layer of the generator. This regularizer penalizes the distances between the first layer's output of the paired input, where the paired input should consist of identical latent variables but different domain variables. As a result, it enforces the generator to decode similar high-level semantics for the paired input, because the first layer decodes the highest-level semantics. This strategy is based on the fact that corresponding images from different domains always share some high-level semantics. As shown in Fig. 2.18c and d, the second regularizer is added to the last hidden layer of the discriminator, which is responsible for encoding the highest-level semantics.

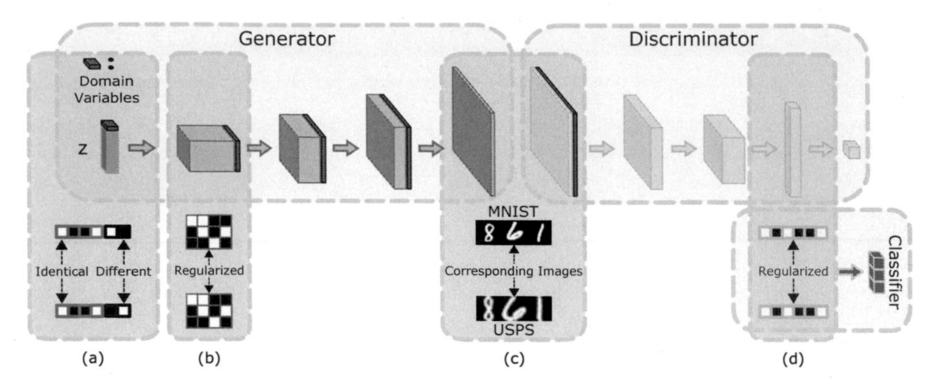

Fig. 2.18 Framework of RCGANs. The domain variables (in purple) are conditioned to all layers of the generator and the input image layer of the discriminator. (**a**): The input pairs consist of identical latent variables but different domain variables. (**b**): One regularizer is used in the first layer of the generator. It penalizes the distances between the first layer's output of the input pairs in (**a**), which guides the generator to decode similar high-level semantics for corresponding images. (**c**): The generator generates pairs of corresponding images. (**d**): Another regularizer is used in the last hidden layer of the discriminator. This regularizer enforces the discriminator to output similar losses for the corresponding images. These similar losses are used to update the generator. This regularizer also makes the model output invariant feature representations for the corresponding images from different domains

This regularizer enforces the discriminator to output similar losses for the pairs of corresponding images. These similar losses are then used to update the generator, which guides the generator to generate similar (corresponding) images.

We model the task of multi-domain image generation to learn a conditional distribution $p_{\text{data}}(\boldsymbol{x}|d)$ over data $\boldsymbol{x}$, where d denotes the domain variable. We propose Regularized Conditional GANs (RCGANs) to learn $p_{\text{data}}(\boldsymbol{x}|d)$. Our idea is to encode the domain-specific semantics in the domain variable d and to encode the common semantics in the shared latent variables z. To achieve this, the conditional GANs are adopted, and two regularizers are proposed. One regularizer is added to the first layer of the generator, and the other is added to the last hidden layer of the discriminator.

Specifically, as Fig. 2.18 shows, for an input pair (zd_i, zd_j) with identical z but different d, the first regularizer penalizes the distance between the first layer's output of zd_i and zd_j, which enforces G to decode similar high-level semantics, since the first layer decodes the highest-level semantics. In contrast, for a pair of corresponding images $(\boldsymbol{x}_i, \boldsymbol{x}_j)$, the second regularizer penalizes the distance between the last layer's output of $\boldsymbol{x}_i$ and $\boldsymbol{x}_j$. As a result, D outputs similar losses for the pairs of corresponding images. When updating G, these similar losses guide G to generate similar (corresponding) images. Note that to use the above two regularizers requires constructing pairs of input with identical z but different d.

Formally, when training, we construct mini-batches with pairs of input $(z, d = 0)$ and $(z, d = 1)$, where the noise input z is the same. G maps the noise input z to a conditional data space $G(z|d)$. An L2-norm regularizer is used to enforce $G_{h_0}(z|d)$, the output of G's first layer, to be similar for each paired input. Another L2-norm regularizer is used to enforce $D_{h_i}(G(z|d))$, the output of D's last hidden layer, to be similar for each paired input. The objective function of RCGANs can then be formulated as follows:

$$\min_G \max_D V\ (G, D) = \mathcal{L}_{\text{GAN}}(G, D) + \lambda \mathcal{L}_{\text{reg}}(G) + \beta \mathcal{L}_{\text{reg}}(D),$$

$$\mathcal{L}_{\text{GAN}}(G, D) = \mathbb{E}_{x \sim p_{\text{data}}(x|d)}[\log D(x|d)] + \mathbb{E}_{z \sim p_z(z)}[\log(1 - D(G(z|d)))],$$

$$\mathcal{L}_{\text{reg}}(G) = \mathbb{E}_{z \sim p_z(z)}[\|G_{h_0}(z|d=0) - G_{h_0}(z|d=1)\|^2],$$

$$\mathcal{L}_{\text{reg}}(D) = \mathbb{E}_{z \sim p_z(z)}[-\|D_{h_i}(G(z|d=0)) - D_{h_i}(G(z|d=1))\|^2], \tag{2.18}$$

where the scalars λ and β are used to adjust the weights of the regularization terms; $\|\cdot\|$ denotes the l^2-norm; $d = 0$ and $d = 1$ denote the source domain and target domain, respectively; $G_{h_0}(\cdot)$ denotes the output of G's first layer; and $D_{h_i}(\cdot)$ denotes the output of D's last hidden layer.

2.4.1 Experiments

2.4.1.1 Implementation Details

Except for the tasks about digits (i.e., MNIST and USPS), we adopt LSGANs (Mao et al. 2017) to train the models because LSGANs generate higher-quality images and show greater stability. For digits tasks, we still adopt standard GANs because we find that LSGANs will sometimes generate unaligned digit pairs.

We use Adam optimizer with learning rates of 0.0005 for LSGANs and 0.0002 for standard GANs. For the hyper-parameters in Equations 2.18 and 3.8, we set $\lambda = 0.1$, $\beta = 0.004$, and $\gamma = 1.0$ found by grid search.

2.4.1.2 Digits

We first evaluate RCGANs on MNIST and USPS datasets. Because the image sizes of MNIST and USPS differ, we resize the images in USPS to the same resolution (i.e., 28×28) of MNIST. We train RCGANs for the following three tasks. Following literature (Liu and Tuzel 2016), the first two tasks are to generate the (1) digits and edge digits and (2) digits and negative digits. The third is to generate MNIST and USPS digits. For these tasks, we design the network architecture following the suggestions in literature (Radford et al. 2015), where the generator consists of four transposed convolutional layers and the discriminator is a variant of LeNet (Lecun et al. 1998). The generated image pairs are shown in Fig. 2.19, which clearly shows that RCGANs succeed to generate corresponding digits for all three tasks.

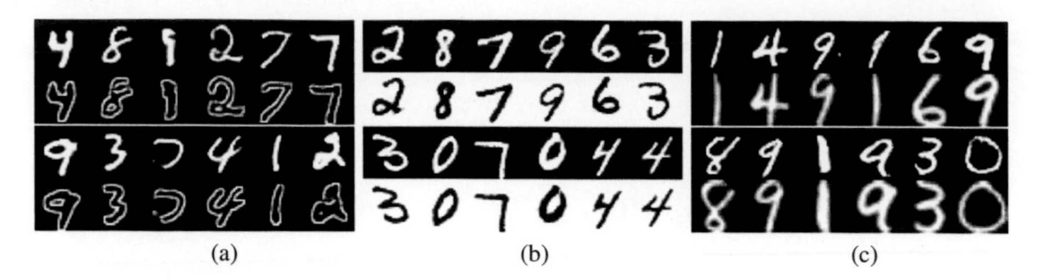

Fig. 2.19 Generated image pairs on digits. (a) Digits and edge digits. (b) Digits and negative digits. (c) MNIST and USPS digits

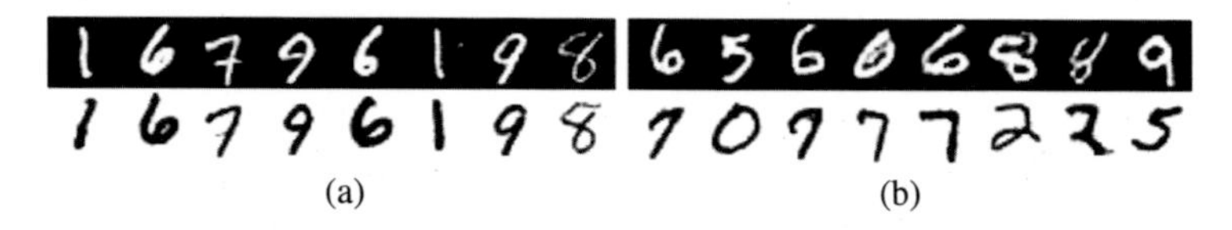

Fig. 2.20 Comparison experiments between models with and without the regularizer. (a) With regularizer. (b) Without regularizer

Without Regularizer

If we remove the proposed regularizers in RCGANs, the model will fail to generate corresponding digits, as Fig. 2.20 shows. This demonstrates that the proposed regularizers play an important role in generating corresponding images.

2.4.1.3 Edges and Photos

We also train RCGANs for the task of generating corresponding edges and photos. The Handbag (Zhu et al. 2016) and Shoe (Yu and Grauman 2014) datasets are used for this task. We randomly shuffle the edge images and realistic photos to avoid using the pair information. We resize all images to a resolution of 64×64. For the network architecture, both the generator and the discriminator consist of four transposed/strided convolutional layers. As shown in Fig. 2.21a and b, RCGANs can generate corresponding images of edges and photos.

Comparison with CoGANs

We also train CoGANs, which are the current state-of-the-art method, on edges and photos using the official implementation of CoGANs. We evaluate two network architectures for CoGANs: (1) the architecture used in CoGANs (Liu and Tuzel 2016) and (2) the same architecture to RCGANs. We also evaluate the standard GANs loss and least squares loss (LSGANs) for CoGANs. However, none of these settings generate corresponding images of edges and photos. The results are shown in Fig. 2.21c and d.

Fig. 2.21 Generated image pairs on shoes and handbags. (**a**) Shoes generated by RCGANs (Ours). (**b**) Handbags generated by RCGANs (Ours). (**c**) Shoes generated by CoGANs. (**d**) Handbags generated by CoGANs

2.4.1.4 Faces

In this task, we evaluate RCGANs on the CelebA dataset (Liu et al. 2014). We first apply a pre-processing method to crop the facial region in the center of the images (Karras et al. 2017) and then resize all the cropped images to a resolution of 112×112. The network architecture used in this task is similar to that in Sect. 2.4.1.3 except for the output dimensions of the layers. We investigate four tasks: (1) female with different color hairs, (2) male with and without glasses, (3) male and female, and (4) male with and without sideburns. The results are presented in Fig. 2.22. We also show the results of three domains in Fig. 2.23 and the interpolation results in Fig. 2.24. The comparison result with CoGAN is shown in Fig. 2.25. The image pairs generated by RegCGAN are more consistent and have better quality than the ones generated by CoGAN.

Comparison with CoGANs
The image pairs generated by CoGANs are also presented in Fig. 2.25, where the image pairs of black and blond hair by CoGANs are duplicated from Liu and Tuzel (2016). We observe that the image pairs generated by RCGANs are more consistent and of better quality than those generated by CoGANs, especially for the task of female and male, which is more difficult than the task of blond and black hair.

Comparison with CycleGANs
We also compare RCGANs with CycleGANs (Zhu et al. 2017), which are the state-of-the-art method in image-to-image transition. To compare with CycleGANs, we first generate some image pairs using RCGANs and then use the generated images in one domain as the input for CycleGANs. The results are presented in Fig. 2.26. Compared with RCGANs, CycleGANs introduce some blur to the generated images. Moreover, the color of the image pairs generated by RCGANs is more consistent than that of those generated by CycleGANs.

Fig. 2.22 Generated image pairs on faces with different attributes. (**a**): Female with different color hairs. (**b**): Male with and without glasses. (**c**): Male and female. (**d**): Male with and without sideburns

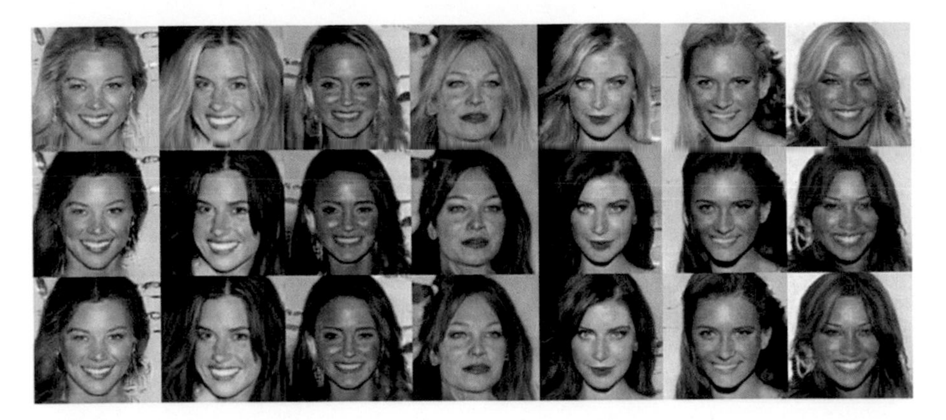

Fig. 2.23 Results of three domains

Fig. 2.24 Interpolation results of female and male

2.4.1.5 Quantitative Evaluation

To further evaluate the effectiveness of RCGANs, we conduct a user study on Amazon Mechanical Turk (AMT). For this evaluation, we also use the task of the female and male generation. In particular, given two image pairs randomly selected from RCGANs and CoGANs, the AMT annotators are asked to choose a better one based on the image quality, perceptual realism, and appearance consistency of female and male. As shown in Table 2.6, with 3000 votes, most annotators preferred the image pairs from RCGANs (77.6%), demonstrating that the overall image quality of our model is better than that of CoGANs.

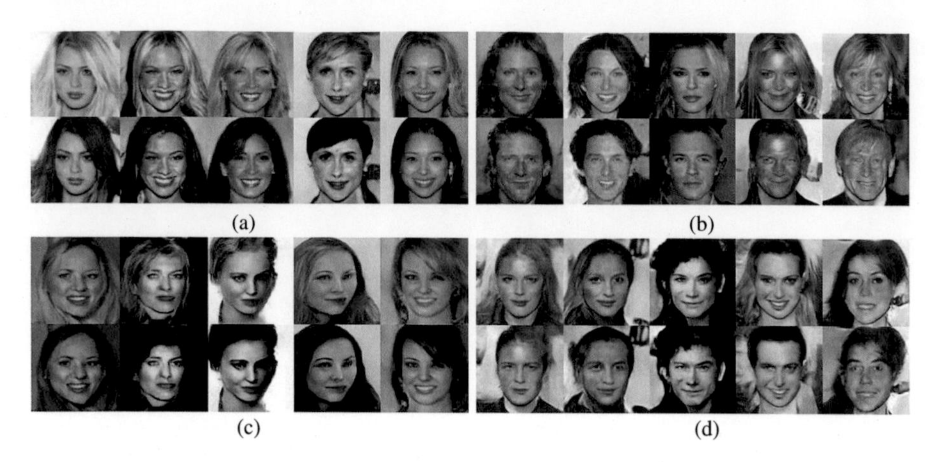

Fig. 2.25 Generated image pairs on faces with different attributes. The image pairs of black and blond hair by CoGANs are duplicated from the CoGANs paper. (**a**) Blond and black hair by RCGANs (Ours). (**b**) Female and male by RCGANs (Ours). (**c**) Blond and black hair by CoGANs. (**d**) Female and male by CoGANs

Fig. 2.26 Comparison results between RCGANs and CycleGANs for the task of female and male. The top two rows are generated by RCGANs. The third row is generated by CycleGANs using the first row as input

Table 2.6 A user study on the task of female and male generation. With 3000 votes, 77.6% of the annotators preferred the image pairs from RCGANs

	CoGANs	RCGANs (Ours)
User choice	673/3000 (22.4%)	**2327**/3000 (**77.6%**)

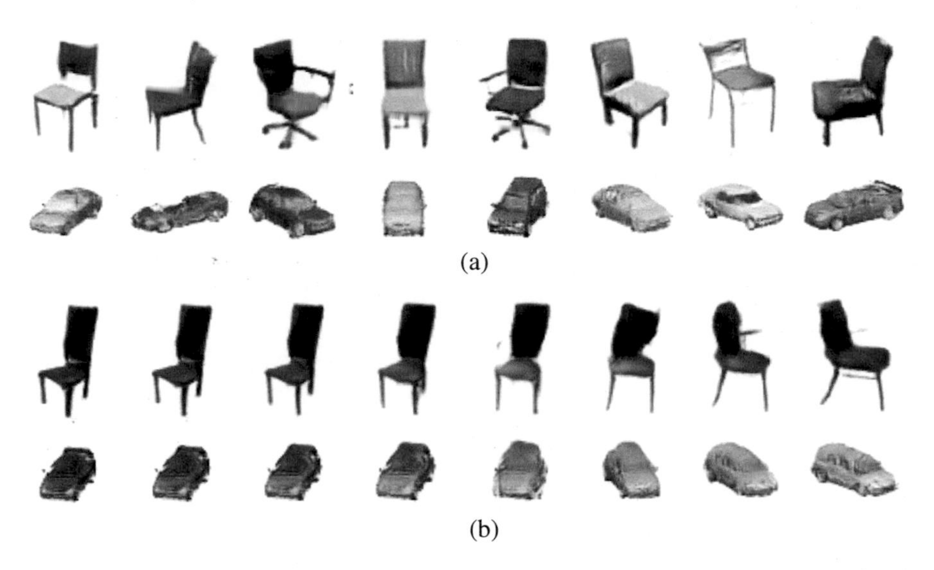

Fig. 2.27 Generated image pairs on chairs and cars, where the orientations are highly correlated. (**a**): Chairs and cars. (**b**): Interpolation between two random points in noise input

2.4.1.6 More Applications

Chairs and Cars
In this task, we use two visually completely different datasets: chairs (Aubry et al. 2014) and cars (Fidler et al. 2012). Both datasets contain synthesized samples with different orientations. We train RCGANs on these two datasets to study whether it can generate corresponding images that share the same orientations. The generated results are shown in Fig. 2.27 with an image resolution of 64 × 64. We further perform interpolation between two random points in the latent space as shown in Fig. 2.27b. The interpolation shows smooth transitions of chairs and cars both in viewpoint and style, while the chairs and cars continue to face the same direction.

Photos and Depths
The NYU depth dataset (Silberman et al. 2012) is used to learn RCGANs over photos and depth images. In this task, we first resize all images to a resolution of 120 × 160 and then randomly crop 112 × 112 patches for training. Figure 2.28 shows the generated image pairs.

Photos and Monet-Style Images
In this task, we train RCGANs on the Monet-style dataset (Zhu et al. 2017). We use the same pre-processing method as in section "Photos and Depths". Figure 2.29 shows the generated image pairs.

Summer and Winter
We also train RCGANs on the summer and winter dataset (Zhu et al. 2017). We use the same pre-processing method as in section "Photos and Depths". Figure 2.30 shows the generated image pairs.

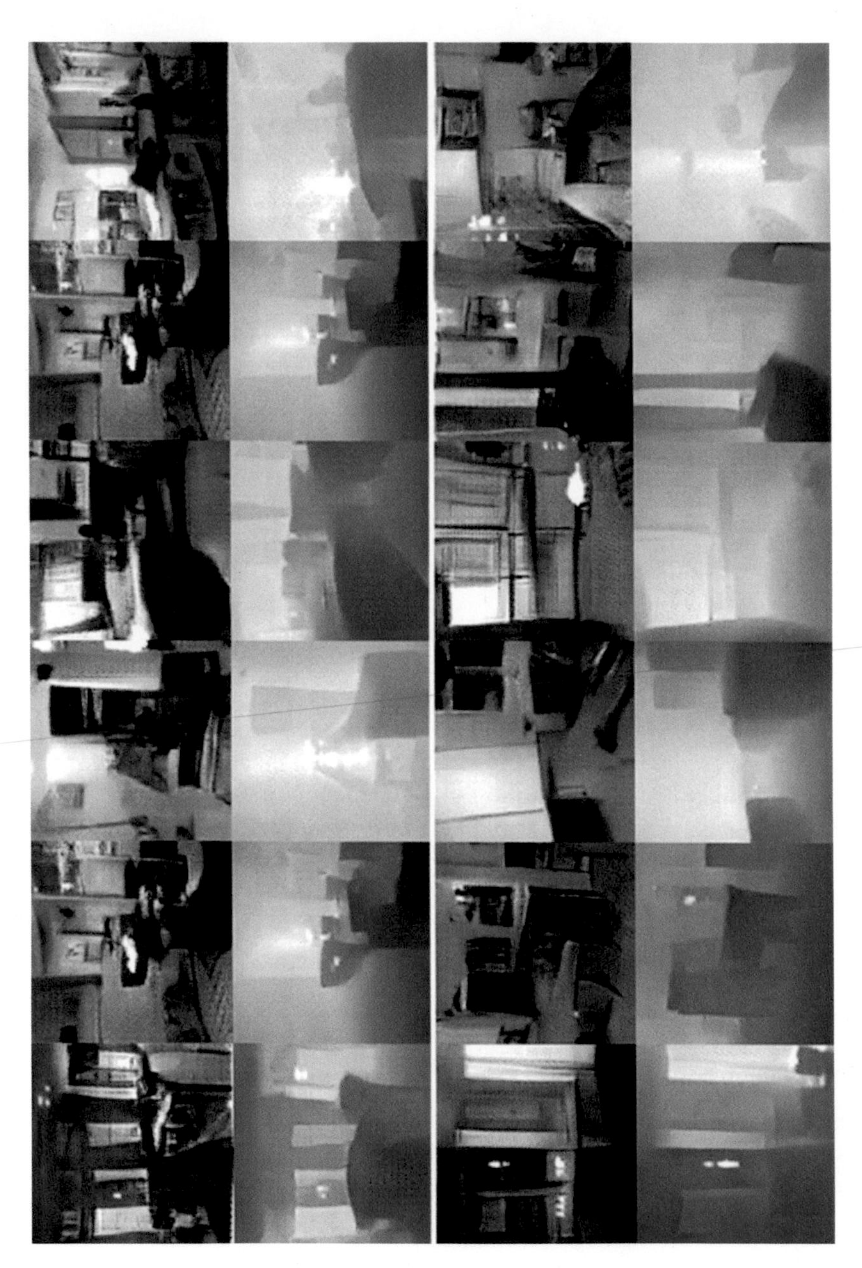

Fig. 2.28 Generated image pairs on photos and depth images

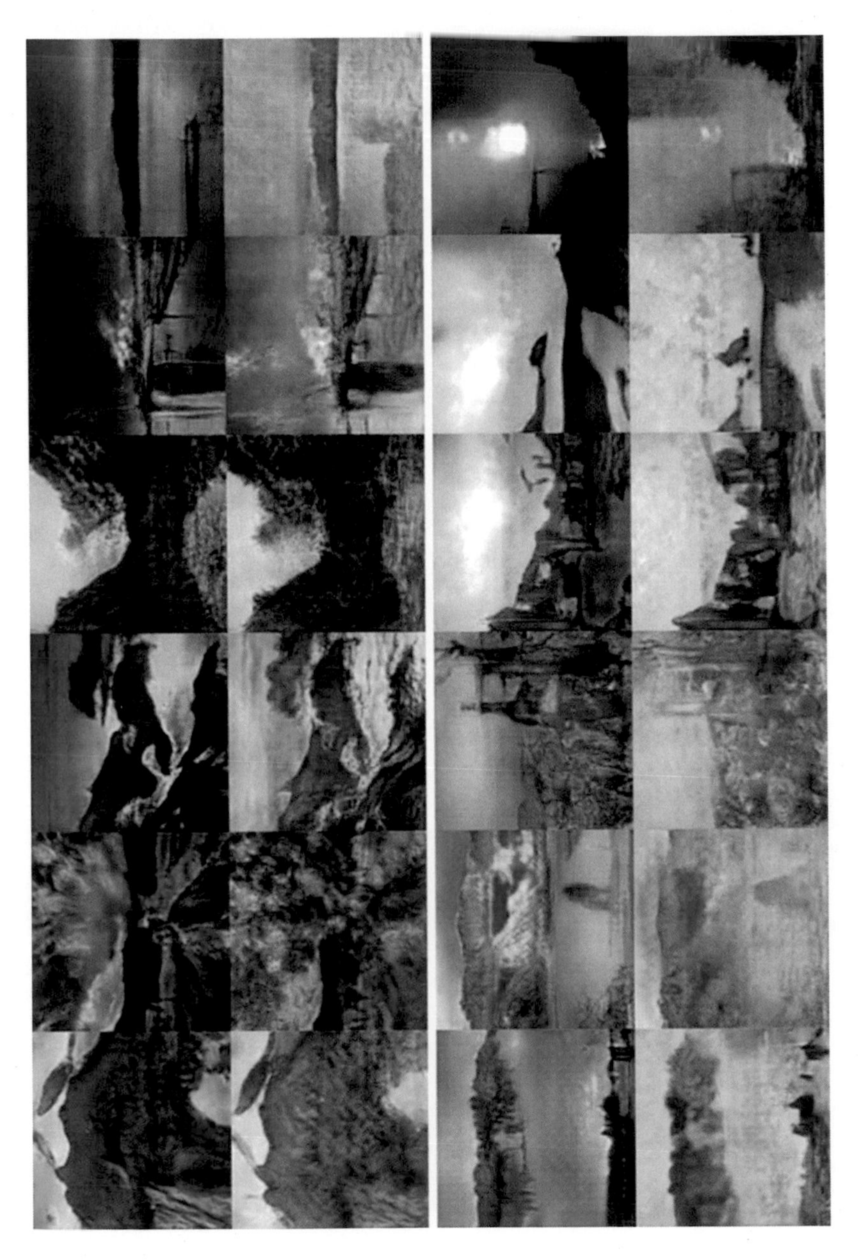

Fig. 2.29 Generated image pairs on photos and Monet-style images

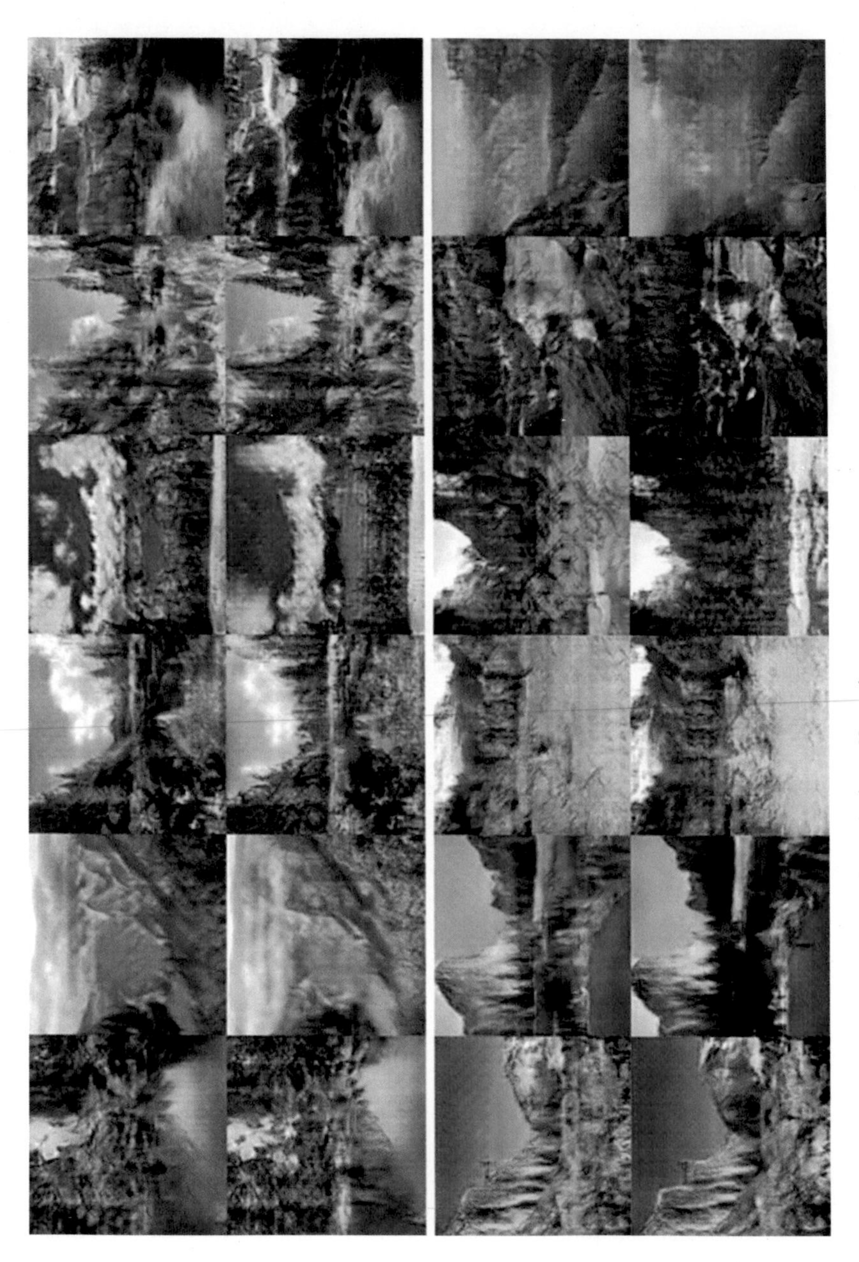

Fig. 2.30 Generated image pairs on summer Yosemite and winter Yosemite

Bibliography

Abadi M, Agarwal A, Barham P, Brevdo E, Chen Z, Citro C, Corrado GS, Davis A, Dean J, Devin M, Ghemawat S, Goodfellow IJ, Harp A, Irving G, Isard M, Jia Y, Józefowicz R, Kaiser L, Kudlur M, Levenberg J, Mané D, Monga R, Moore S, Murray DG, Olah C, Schuster M, Shlens J, Steiner B, Sutskever I, Talwar K, Tucker PA, Vanhoucke V, Vasudevan V, Viégas FB, Vinyals O, Warden P, Wattenberg M, Wicke M, Yu Y, Zheng X (2016) TensorFlow: large-scale machine learning on heterogeneous distributed systems. arXiv:1603.04467

Agustsson E, Tschannen M, Mentzer F, Timofte R, Gool LV (2018) Generative adversarial networks for extreme learned image compression. arXiv:1804.02958

Arjovsky M, Bottou L (2017) Towards principled methods for training generative adversarial networks. arXiv:1701.04862

Arjovsky M, Chintala S, Bottou L (2017) Wasserstein GAN. In: International conference on machine learning (ICML), pp 214–223

Arora S, Ge R, Liang Y, Ma T, Zhang Y (2017) Generalization and equilibrium in generative adversarial nets (GANs). arXiv:1703.00573

Aubry M, Maturana D, Efros A, Russell B, Sivic J (2014) Seeing 3D chairs: exemplar part-based 2D-3D alignment using a large dataset of CAD models. In: Computer vision and pattern recognition (CVPR)

Berthelot D, Schumm T, Metz L (2017) BEGAN: boundary equilibrium generative adversarial networks. arXiv:1703.10717

Bishop CM (2006) Pattern recognition and machine learning

Brock A, Donahue J, Simonyan K (2018) Large scale GAN training for high fidelity natural image synthesis. arXiv:1809.11096

Che T, Li Y, Jacob AP, Bengio Y, Li W (2016) Mode regularized generative adversarial networks. arXiv:1612.02136

Chen X, Duan Y, Houthooft R, Schulman J, Sutskever I, Abbeel P (2016) InfoGAN: interpretable representation learning by information maximizing generative adversarial nets. In: Advances in neural information processing systems (NeurIPS), pp 2172–2180

Dai Z, Almahairi A, Bachman P, Hovy EH, Courville AC (2017) Calibrating energy-based generative adversarial networks. arXiv:1702.01691

Denton E, Chintala S, Szlam A, Fergus R (2015) Deep generative image models using a Laplacian pyramid of adversarial networks. In: Advances in neural information processing systems (NeurIPS), pp 1486–1494

Dieng AB, Tran D, Ranganath R, Paisley J, Blei DM (2017) Variational inference via χ upper bound minimization. In: Advances in neural information processing systems (NeurIPS), pp 2732–2741

Dinh L, Krueger D, Bengio Y (2014) NICE: non-linear independent components estimation. arXiv:1410.8516

Donahue J, Krähenbühl P, Darrell T (2016) Adversarial feature learning. arXiv:1605.09782

Dosovitskiy A, Springenberg JT, Brox T (2015) Learning to generate chairs, tables and cars with convolutional networks. In: Computer vision and pattern recognition (CVPR)

Dumoulin V, Belghazi I, Poole B, Lamb A, Arjovsky M, Mastropietro O, Courville AC (2016) Adversarially learned inference. arXiv:1606.00704

Fidler S, Dickinson S, Urtasun R (2012) 3D object detection and viewpoint estimation with a deformable 3D cuboid model. In: Advances in neural information processing systems (NeurIPS)

Ganin Y, Ustinova E, Ajakan H, Germain P, Larochelle H, Laviolette F, Marchand M, Lempitsky V (2016) Domain-adversarial training of neural networks. J Mach Learn Res 17(1):2096–2030

Goodfellow I, Pouget-Abadie J, Mirza M, Xu B, Warde-Farley D, Ozair S, Courville A, Bengio Y (2014) Generative adversarial nets. In: Advances in neural information processing systems (NeurIPS), pp 2672–2680

Goodfellow I, Bengio Y, Courville A (2016) Deep learning. MIT Press, Cambridge, MA

Gulrajani I, Ahmed F, Arjovsky M, Dumoulin V, Courville A (2017) Improved training of Wasserstein GANs. In: Advances in neural information processing systems (NeurIPS), pp 5767–5777

He K, Zhang X, Ren S, Sun J (2016) Deep residual learning for image recognition. In: Computer vision and pattern recognition (CVPR), pp 770–778

Heusel M, Ramsauer H, Unterthiner T, Nessler B, Hochreiter S (2017) GANs trained by a two time-scale update rule converge to a local nash equilibrium. In: Advances in neural information processing systems (NeurIPS), pp 6626–6637

Hinton G, Salakhutdinov R (2006) Reducing the dimensionality of data with neural networks. Science 313(5786):504–507

Hinton GE Salakhutdinov RR (2009) Replicated softmax: an undirected topic model. In: Advances in neural information processing systems (NeurIPS), pp 1607–1614

Hinton GE, Osindero S, Teh Y-W (2006) A fast learning algorithm for deep belief nets. Neural Comput 18(7):1527–1554

Huang X, Li Y, Poursaeed O, Hopcroft J, Belongie S (2017) Stacked generative adversarial networks. In: Computer vision and pattern recognition (CVPR), pp 5077–5086

Huszar F (2015) How (not) to train your generative model: scheduled sampling, likelihood, adversary? arXiv:1511.05101

Isola P, Zhu J-Y, Zhou T, Efros AA (2017) Image-to-image translation with conditional adversarial networks. In: Computer vision and pattern recognition (CVPR), pp 5967–5976

Jolicoeur-Martineau A (2018) The relativistic discriminator: a key element missing from standard GAN. arXiv:1807.00734

Karras T, Aila T, Laine S, Lehtinen J (2017) Progressive growing of GANs for improved quality, stability, and variation. arXiv:1710.10196

Kingma DP, Ba J (2014) Adam: a method for stochastic optimization. arXiv:1412.6980

Kingma DP, Welling M (2013) Auto-encoding variational Bayes. arXiv:1312.6114

Kodali N, Abernethy JD, Hays J, Kira Z (2017) On convergence and stability of GANs. arXiv:1705.07215

Larsen ABL, Sønderby SK, Winther O (2015) Autoencoding beyond pixels using a learned similarity metric. arXiv:1512.09300

Lecun Y, Bottou L, Bengio Y, Haffner P (1998) Gradient-based learning applied to document recognition. In: Proceedings of the IEEE, pp 2278–2324

Ledig C, Theis L, Huszar F, Caballero J, Cunningham A, Acosta A, Aitken A, Tejani A, Totz J, Wang Z, Shi W (2017) Photo-realistic single image super-resolution using a generative adversarial network. In: Computer vision and pattern recognition (CVPR), pp 4681–4690

Li C, Liu H, Chen C, Pu Y, Chen L, Henao R, Carin L (2017) ALICE: towards understanding adversarial learning for joint distribution matching. In: Advances in neural information processing systems (NeurIPS), pp 5495–5503

Liu M-Y, Tuzel O (2016) Coupled generative adversarial networks. In: Advances in neural information processing systems (NeurIPS)

Liu Z, Luo P, Wang X, Tang X (2014) Deep learning face attributes in the wild. arXiv:1411.7766

Long J, Shelhamer E, Darrell T (2015) Fully convolutional networks for semantic segmentation. In: Computer vision and pattern recognition (CVPR), pp 3431–3440

Makhzani A, Shlens J, Jaitly N, Goodfellow IJ (2015) Adversarial autoencoders. arXiv:1511.05644

Mao X, Li Q, Xie H, Lau RY, Wang Z, Smolley SP (2017) Least squares generative adversarial networks. In: International conference on computer vision (ICCV), pp 2794–2802

Mathieu M, Couprie C, LeCun Y (2015) Deep multi-scale video prediction beyond mean square error. arXiv:1511.05440

Mescheder L, Nowozin S, Geiger A (2017a) The numerics of GANs. In: Advances in neural information processing systems (NeurIPS), pp 1825–1835

Mescheder LM, Nowozin S, Geiger A (2017b) Adversarial variational Bayes: unifying variational autoencoders and generative adversarial networks. arXiv:1701.04722

Metz L, Poole B, Pfau D, Sohl-Dickstein J (2016) Unrolled generative adversarial networks. arXiv:1611.02163

Mirza M, Osindero S (2014) Conditional generative adversarial nets. arXiv:1411.1784

Netzer Y, Wang T, Coates A, Bissacco A, Wu B, Ng AY (2011) Reading digits in natural images with unsupervised feature learning. In: NeurIPS workshop on deep learning and unsupervised feature learning

Nguyen X, Wainwright MJ, Jordan MI (2010) Estimating divergence functionals and the likelihood ratio by convex risk minimization. In: IEEE Trans Inf Theory 56(11):5847–5861

Nguyen A, Yosinski J, Bengio Y, Dosovitskiy A, Clune J (2017) Plug & play generative networks: conditional iterative generation of images in latent space. In: Computer vision and pattern recognition (CVPR), pp 4467–4477

Nowozin S, Cseke B, Tomioka R (2016) f-GAN: training generative neural samplers using variational divergence minimization. In: Advances in neural information processing systems (NeurIPS), pp 271–279

Odena A, Olah C, Shlens J (2016) Conditional image synthesis with auxiliary classifier GANs. arXiv:1610.09585

Perarnau G, van de Weijer J, Raducanu B, Álvarez JM (2016) Invertible conditional GANs for image editing. arXiv:1611.06355

Qi G-J (2017) Loss-sensitive generative adversarial networks on Lipschitz densities. arXiv:1701.06264

Radford A, Metz L, Chintala S (2015) Unsupervised representation learning with deep convolutional generative adversarial networks. arXiv:1511.06434

Reed S, Akata Z, Yan X, Logeswaran L, Schiele B, Lee H (2016) Generative adversarial text-to-image synthesis. In: International conference on machine learning (ICML), pp 1060–1069

Ren S, He K, Girshick R, Sun J (2015) Faster R-CNN: towards real-time object detection with region proposal networks. In: Advances in neural information processing systems (NeurIPS), pp 91–99

Roth K, Lucchi A, Nowozin S, Hofmann T (2017) Stabilizing training of generative adversarial networks through regularization. In: Advances in neural information processing systems (NeurIPS), pp 2018–2028

Salakhutdinov R, Hinton G (2009) Deep Boltzmann machines. In: International conference on artificial intelligence and statistics, pp 448–455

Salimans T, Goodfellow I, Zaremba W, Cheung V, Radford A, Chen X, Chen X (2016) Improved techniques for training GANs. In: Advances in neural information processing systems (NeurIPS), pp 2226–2234

Shrivastava A, Pfister T, Tuzel O, Susskind J, Wang W, Webb R (2016) Learning from simulated and unsupervised images through adversarial training. arXiv:1612.07828

Silberman N, Hoiem D, Kohli P, Fergus R (2012) Indoor segmentation and support inference from RGBD images. In: European conference on computer vision (ECCV)

Springenberg JT (2018) Unsupervised and semi-supervised learning with categorical generative adversarial networks. arXiv:1511.06390

Taigman Y, Polyak A, Wolf L (2016) Unsupervised cross-domain image generation. arXiv:1611.02200

Taylor GW, Fergus R, LeCun Y, Bregler C (2010) Convolutional learning of spatio-temporal features. In: European conference on computer vision (ECCV), pp 140–153

Tieleman T, Hinton G (2012) Lecture 6.5lRMSProp: divide the gradient by a running average of its recent magnitude. In: COURSERA: neural networks for machine learning

Vondrick C, Pirsiavash H, Torralba A (2016) Generating videos with scene dynamics. arXiv:1609.02612

Wang X, Gupta A (2016) Generative image modeling using style and structure adversarial networks. In: European conference on computer vision (ECCV)

Wang C, Xu C, Tao D (2017a) Tag disentangled generative adversarial networks for object image re-rendering. In: International joint conference on artificial intelligence (IJCAI)

Wang T-C, Liu M-Y, Zhu J-Y, Tao A, Kautz J, Catanzaro B (2017b) High-resolution image synthesis and semantic manipulation with conditional GANs. arXiv:1711.11585
Wang T-C Liu M-Y, Zhu J-Y, Liu G, Tao A, Kautz J, Catanzaro B (2018) Video-to-video synthesis. arXiv:1808.06601
Yan X, Yang J, Sohn K, Lee H (2015) Attribute2Image: conditional image generation from visual attributes. arXiv:1512.00570
Yu A, Grauman K (2014) Fine-grained visual comparisons with local learning. In: Computer vision and pattern recognition (CVPR)
Zhang W, Sun J, Tang X (2008) Cat head detection – how to effectively exploit shape and texture features. In: European conference on computer vision (ECCV), pp 802–816
Zhang H, Goodfellow I, Metaxas D, Odena A (2018) Self-attention generative adversarial networks. arXiv:1805.08318
Zhao JJ, Mathieu M, LeCun Y (2016) Energy-based generative adversarial network. arXiv:1609.03126
Zhu J-Y, Krähenbühl P, Shechtman E, Efros AA (2016) Generative visual manipulation on the natural image manifold. In: European conference on computer vision (ECCV). arXiv:1609.03552
Zhu J-Y, Park T, Isola P, Efros AA (2017) Unpaired image-to-image translation using cycle-consistent adversarial networks. In: International conference on computer vision (ICCV)

Chapter 3
More Key Applications of GANs

In the previous chapter, we learned that GANs are very powerful for image generation. In this chapter, we learn three interesting applications of GANs: image-to-image translation, unsupervised domain adaptation, and GANs for security. One type of GANs application is for tasks that require high-quality images, such as image-to-image translation. To improve the output image quality, a discriminator is introduced to judge whether the output images are realistic. Another type is to extend the use of the discriminator in a more generalized way, such as unsupervised domain adaptation. For instance, the task of unsupervised domain adaptation includes two domain datasets, the source domain and the target domain. To learn indistinguishable feature representations for the source and target domains, we can introduce a discriminator to judge whether the output features come from the source domain or the target domain.

3.1 Image-to-Image Translation

Image-to-image translation is among the most successful applications of GANs. Image-to-image translation is a general framework that translates an input image into a corresponding output image. Many vision and graphics tasks can be involved, such as semantic segmentation, image colorization, and style transfer. In this section, we introduce two famous image-to-image translation models, pix2pix (Isola et al. 2017) and CycleGANs (Zhu et al. 2017). In the previous chapters, the generator always starts with a noise vector. For image-to-image translation, the generator changes to start with an image, and thus the architecture of the generator becomes an encoder-decoder framework. Generally, the image-to-image translation task has two requirements for the output images: (1) the output images should look realistic from the target domain, and (2) there should be semantic consistency between the input and output images. The first requirement can be achieved by

X. Mao, Q. Li, *Generative Adversarial Networks for Image Generation*,
https://doi.org/10.1007/978-981-33-6048-8_3

using a discriminator to distinguish whether the output images come from the target domain or from the generator. Therefore, the critical problem for the image-to-image translation problem is how to constrain the semantic consistency between the input and output images. To this end, the pix2pix model used the conditional GAN, and the CycleGANs model used the cycle-consistency constraint. The details of pix2pix and CycleGANs are introduced in the following sections, and their variants are also presented.

3.1.1 pix2pix

The pix2pix (Isola et al. 2017) model was the first to define image-to-image translation as the problem of translating one domain image into another domain image, such that many traditional tasks, such as photographs to edges and semantic labels to realistic images, can be involved in image-to-image translation. Note that the pix2pix model is designed for paired training data, for which the pairs of input and output images are well aligned. In contrast, the CycleGANs model introduced in the next section is designed for unpaired training data.

As mentioned above, image-to-image translation has two requirements for the output images: (1) the output images should look realistic from the target domain, and (2) there should be semantic consistency between the input and output images. For the first requirement, the pix2pix model uses GANs to make the output images look realistic from the target domain. Specifically, the discriminator is trained to distinguish whether the output images are real from the target domain, such that the generator can output images that look realistic from the target domain. For the second requirement, the pix2pix model uses conditional GANs (cGANs) to constrain the semantic consistency between the input and output images. The framework of the pix2pix model is shown in Fig. 3.1. Specifically, both the generator and discriminator are conditioned on the input images. The input of the discriminator is a pair of input and output images. The discriminator treats the pair of the input image and its corresponding ground-truth image as real data and treats the pair of the input image and the synthesized output image as fake data. As the real image pairs are well aligned, the generator can fool the discriminator only if it can output images that are well aligned to the input images. This is the intuition behind the use of cGANs to constrain the semantic consistency between the input and output images.

Formally, the generator G of cGANs maps an input image x and a random noise vector z to an output image y. The discriminator D maps a pair of input and output images to a scalar value that indicates whether the image pair is real or fake. The generator G is trained adversarially to produce images that cannot be distinguished

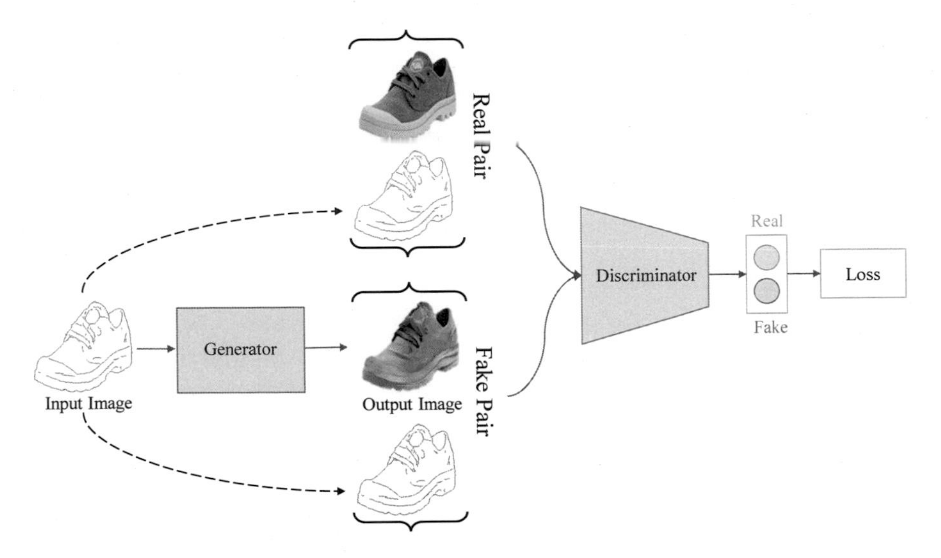

Fig. 3.1 Framework of pix2pix

from real images by the discriminator D. The objective of these cGANs can be defined as

$$\min_G \max_D \mathcal{L}(G, D) = \mathbb{E}_{(x,y)}[\log D(y|x)] + \mathbb{E}_{x,z}[\log(1 - D(G(z|x)|x))], \tag{3.1}$$

where (x, y) is an aligned image pair from the training data.

In addition, because the ground-truth images are provided for the input images, we can apply a loss to constrain the pixel distance between the generated and ground-truth images. The pix2pix model uses L1 loss for this:

$$\min_G \mathcal{L}_{L1}(G) = \mathbb{E}_{(x,y),z}[||y - G(z|x)||_1], \tag{3.2}$$

where (x, y) is an aligned image pair from the training data.

The intention behind the use of the noise vector z is to introduce the stochasticity for the output of the generator. However, in practice, the generator learns to ignore the noise vector, so the pix2pix model removes the noise vector in practice. Therefore, the learned mapping of pix2pix is deterministic and one-to-one. The development of a one-to-many mapping method remains an open problem.

For the network architectures, the pix2pix model uses U-Net (Ronneberger et al. 2015) for the generator. Specifically, the architecture of the generator is an encoder-decoder network with skip connections between corresponding layers in the encoder and decoder. The architecture of the discriminator is a patch-based network. Specifically, the discriminator is a fully convolutional network that outputs $N \times N$ responses that correspond to the $N \times N$ patches of the image. Therefore, the discriminator distinguishes whether each patch is real or fake. The intuition of this design is that Equation 3.2 (i.e., L1 loss) is used to restrict the global correctness and the patch-based discriminator is used to restrict the local correctness.

Next, we introduce three variants of the pix2pix model and discuss ideas to improve the performance of pix2pix.

The pix2pixHD model (Wang et al. 2017) extends the pix2pix model for synthesis of 2048 $\times$ 1024 images. The idea of the pix2pixHD model is to use a coarse-to-fine generator and multi-scale discriminators. The coarse-to-fine method can also be found in other GANs models (Denton et al. 2015; Huang et al. 2017; Zhang et al. 2016). In pix2pixHD, the first generator is trained on lower-resolution images, and it is an encoder-decoder network with a set of residual blocks. The second generator is also an encoder-decoder network with a set of residual blocks and is stacked at the first generator. The two generators are trained jointly on high-resolution images. In particular, the last feature map of the second generator's encoder and the last feature map of the first generator's decoder are added in an element-wise manner as the input of the second generator's decoder. Moreover, multi-scale discriminators are used to distinguish high-resolution real and fake images. The discriminator operating at the coarsest scale helps the generator to generate globally realistic images, whereas the discriminator operating at the finest scale guides the generator to refine the details of the generated images.

In pix2pix, the discriminator is conditioned on the input images and is used to constrain the semantic consistency between the input and output images. If the conditioning is removed from the discriminator, the generator will only care whether the output images look realistic. The Perceptual Adversarial Network (PAN) (Wang et al. 2018) uses the perceptual adversarial loss to constrain the semantic consistency between the input and output images, instead of conditioning the discriminator. In particular, the hidden layers of the discriminator are used to evaluate the perceptual adversarial loss between the output images and ground-truth images. The architecture of PAN's generator is similar to that of pix2pix, which is also an encoder-decoder network. In addition, experiments have shown that combining the perceptual adversarial loss with the conditional discriminator does not further improve the performance.

The Deep Attention Generative Adversarial Networks (DA-GANs) (Ma et al. 2018) constrain the instance-level consistency between the input and output images with the use of a deep attention encoder. Unlike PAN (Wang et al. 2018) which constrains semantic consistency on the global information, DA-GANs focus on the instance-level semantic consistency. Specifically, DA-GANs use a deep attention encoder to extract the instance-level attention regions first and then adopt a consistency loss to constrain the discrepancy between the instance-level representations of the input and output images.

The batch normalization layer is usually adopted in the architectures of GANs. However, the batch normalization layer will lead to the problem of eliminating some semantic information from the input images, especially when translating semantic label images to photorealistic images. To address this issue, a conditional normalization method called SPatially-Adaptive (DE)normalization (SPADE) was proposed by Park et al. (2019). Specifically, the input images are encoded into spatially adaptive tensors, and the tensors are multiplied and added to the batch normalized activations in an element-wise manner. In this way, the semantic information of the input images is reserved. Because the input images have been encoded in normalization layers, SPADE removes the encoder network in the generator. For the discriminator architecture, SPADE adopts the same multi-scale discriminator used in pix2pixHD (Wang et al. 2017).

In pix2pix, the use of the noise vector cannot introduce the stochasticity for the output image because the noise vector will be ignored during training. One method to tackle this problem is to force the noise vector to be "useful," as proposed in BicycleGANs (Zhu et al. 2017). BicycleGANs introduce two methods to force the noise vector to be "useful." The first is to use an encoder to map the ground-truth image to the noise vector, which is then used as the input of the generator. The second is to use another encoder to recover the noise vector from the generated image.

3.1.2 CycleGAN

The pix2pix model uses conditional GANs to constrain the semantic consistency between the input and output images. This method assumes that the paired information between the input and output images is provided; however, in many real-world tasks, paired training data are not always available. To this end, the CycleGANs model is proposed for unpaired image-to-image translation.

CycleGANs use the cycle-consistency constraint to impose semantic consistency between the input and output images. Specifically, if we translate an image from domain X to domain Y and then back to domain X, we should arrive back at the original image. The intuition of this method is to reduce the space of possible mappings. If we only apply the translation from domain X to domain Y, there are infinite possible mappings for domain Y. If we add the cycle-consistency constraint, translating from domain X to domain Y is a one-to-many mapping, and translating from domain Y back to domain X is a many-to-one mapping. The most convenient way for the model to learn is to make the translation a one-to-one mapping and to maintain semantic consistency before and after translation.

The framework of CycleGANs is shown in Fig. 3.2. Formally, the goal is to learn translations between two domains X and Y. The model consists of two generators G_X and G_Y and two discriminators D_X and D_Y. Generator G_X maps an image x from domain X to an image $G_X(x)$ from domain Y, and generator G_Y performs in the opposite direction. Discriminator D_X distinguishes between real images x and translated images $G_Y(y)$, and discriminator D_X distinguishes between real images y and translated images $G_X(x)$. The objective of CycleGANs contains two losses, the adversarial loss and the cycle-consistency loss. The adversarial loss is applied to

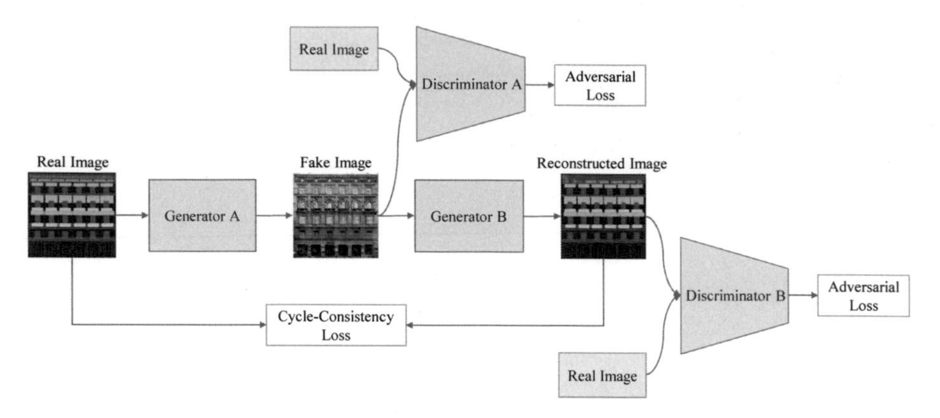

Fig. 3.2 Framework of CycleGANs

both generators G_X and G_Y as

$$\begin{aligned}\min_{G_X}\max_{D_Y}\mathcal{L}_{\text{GAN}}(G_X, D_Y) &= \mathbb{E}_y[\log D_Y(y)] + \mathbb{E}_x[\log(1 - D_Y(G_X(x)))],\\ \min_{G_Y}\max_{D_X}\mathcal{L}_{\text{GAN}}(G_Y, D_X) &= \mathbb{E}_x[\log D_X(x)] + \mathbb{E}_y[\log(1 - D_X(G_Y(y)))].\end{aligned} \tag{3.3}$$

The cycle consistency states that each image x from domain X mapped through the translation cycle should go back to the original image, i.e., $x \to G_X(x) \to G_Y(G_X(x)) \approx x$. Similarly, starting from y, another translation cycle is $y \to G_Y(y) \to G_X(G_Y(y)) \approx y$. CycleGANs adopt the L1 norm to measure the distance between $G_Y(G_X(x))$ and x. The cycle-consistency loss can then be expressed as

$$\min_{G_X,G_Y}\mathcal{L}_{\text{cyc}}(G_X, G_Y) = \mathbb{E}_x[||G_Y(G_X(x)) - x||_1] + \mathbb{E}_y[||G_X(G_Y(y)) - y||_1]. \tag{3.4}$$

The full objective of CycleGANs is

$$\begin{aligned}\min_{G_X,G_Y}\max_{D_X,D_Y}\mathcal{L}(G_X, G_Y, D_X, D_Y) =&\mathcal{L}_{\text{GAN}}(G_X, D_Y) + \mathcal{L}_{\text{GAN}}(G_Y, D_X)\\ &+ \lambda\mathcal{L}_{\text{cyc}}(G_X, G_Y),\end{aligned} \tag{3.5}$$

where λ controls the weight of the cycle-consistency loss.

For the network architectures, CycleGANs adapt the architecture from Johnson et al. (2016) for the generator, which contains two stride-2 convolutions, several residual blocks, and two stride-$\frac{1}{2}$ fractionally strided convolutions. For the discriminator, CycleGANs follow the pix2pix (Isola et al. 2017) model to use a patch-based network as described in Sect. 3.1.1.

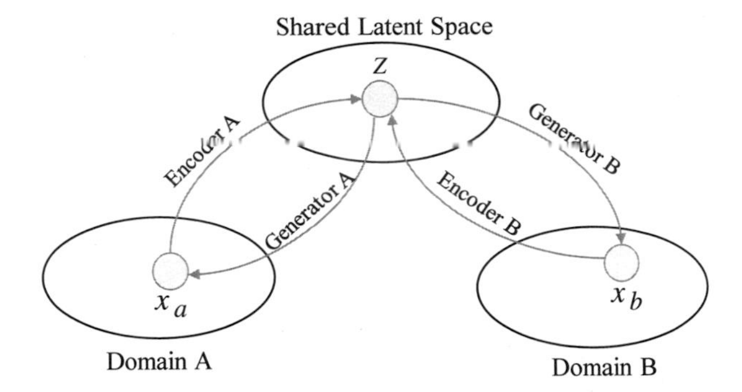

Fig. 3.3 Framework of UNIT. The shared-latent space and the autoencoder framework imply the cycle-consistency constraint ($x_a \rightarrow z \rightarrow x_b \rightarrow z \rightarrow x_a$)

Next, we introduce several variants of the CycleGANs model. As mentioned above, CycleGANs constrain the semantic consistency between the input and output images by using the cycle-consistency loss for unpaired image-to-image translation. We start by introducing different methods to constrain the semantic consistency.

The UNsupervised Image-to-image Translation (UNIT) (Liu et al. 2017) method assumes a shared-latent space for the input and output images, which means that the encoders should output similar latent representations for any pair of input and output images. As Fig. 3.3 shows, the shared-latent space and the autoencoder framework imply the cycle-consistency constraint assumption. UNIT shows that combining the shared-latent space with the cycle-consistency constraint in CycleGANs can achieve the best performance. In practice, the shared-latent space is implemented by enforcing a weight-sharing constraint upon the last few layers of the encoders and the first few layers of the decoders.

The cycle-consistency constraint requires two GANs to translate back the images. The DistanceGANs (Benaim and Wolf 2017) model needs only one GAN to perform image-to-image translation for unpaired data. The intuition of the proposed method called distance constraint is that the distance between two images from the input domain should be preserved in the translation to the output domain. The distance constraint can constrain the semantic consistency between the input and output images without the use of the cycle-consistency constraint. The distance constraint only requires one-sided translation, so only one GAN is used. In the experiments, the proposed distance constraint provides better numerical results than the cycle-consistency constraint, and combining these two constraints usually achieves further improvement.

The Contrastive Unpaired Translation (CUT) (Park et al. 2020) is another method to constrain the semantic consistency between the input and output images. Like the DistanceGAN, CUT is a one-sided translation method for unpaired data. The intuition of CUT is to maximize the mutual information between the input and output image patches using contrastive learning. In particular, given a patch on the output as a "query," the corresponding patch on the input is treated as a "positive" example, and the other patches on the input are treated as "negative" samples; CUT adopts InfoNCE loss (van den Oord et al. 2018) to maximize the mutual information between the "query" and the "positive" example. The relationship between the cycle-consistency constraint and the proposed mutual information constraint is also discussed in the paper (Park et al. 2020). The cycle-consistency loss was shown to be the upper bound of conditional entropy $H(X|Y)$ in Li et al. (2017). Minimizing the cycle-consistency loss is related to the objective of maximizing the mutual information $I(X, Y)$, as $I(X, Y) = H(X) - H(X|Y)$ and $H(X)$ is a constant.

The StarGANs (Choi et al. 2018) model extends CycleGANs to perform image-to-image translation for multiple domains. CycleGANs are designed for one-to-one mapping, and if it is mapping among n domains, it needs to train $n(n-1)$ generators. The method of StarGANs is to attach a domain label to the generator. In particular, given an input image x and a target domain label c, the generator outputs the output image $G(x, c)$. The discriminator is attached with an auxiliary classifier to classify the domain label for the images. In addition to the adversarial loss in GAN, a domain classification loss is introduced for the generator, which makes the generator generate images that can be classified as the target domain.

The ComboGANs (Anoosheh et al. 2018) model is another method to perform image-to-image translation for multiple domains. Each domain has one generator, which consists of an encoder and a decoder, and one discriminator. If it is translating an image from domain X to domain Y, it stacks the encoder of domain X and the decoder of domain Y. For ComboGAN, the number of generators and discriminators scales linearly with the number of domains, instead of n^2 for CycleGAN. Compared with StarGAN, which needs only one generator and one discriminator, ComboGANs need n generators and n discriminators for mapping among n domains. One limitation of StarGANs is that it requires that the images obtained from different domains are sufficiently similar, because StarGANs use a unified generator and a unified discriminator for all domains. ComboGANs do not have this limitation, but it requires more generators and discriminators.

One limitation of CycleGANs is that it is a one-to-one mapping; that is, it outputs a single image for a given input image. A simple method for extending CycleGANs to many-to-many mapping is to attach a noise vector z to the input image. However, in practice, the cycle-consistency constraint makes the model ignore the noise vector z, because when translating back, the model is a many-to-one mapping, and for any z, it translates back to the input image. The Augmented CycleGANs were proposed by Almahairi et al. (2018) to extend the spaces X and Y to augmented spaces $X \times Z_y$ and $Y \times Z_x$. Now, when translating back, the model is not a many-to-one mapping, because for different (y, z_x), the corresponding (x, z_y) also changes. In addition, to simplify the sampling from the noise vector z, it is encouraged to match a simple distribution. Therefore, by sampling different z, the model can output different images for each input image.

The Disentangled Representation for Image-to-Image Translation (DRIT) (Lee et al. 2018) model is another method to perform many-to-many mappings via disentangled representations. Specifically, the representations of images are disentangled into domain-invariant content vectors and domain-specific attribute vectors.

DRIT adopts two techniques to encode the shared content representations: weight sharing and a content discriminator. Weight sharing is similar to the aforementioned UNIT (Liu et al. 2017) model. The content discriminator is used to distinguish the domain membership of the encoded content, and the content encoders adversarially learn to encode domain-invariant representations. In addition, like the aforementioned Augmented CycleGANs (Zhu et al. 2017), the attribute vector is encouraged to match a Gaussian distribution. Therefore, given an input image, by fixing the content vector and sampling different attribute vectors from the target domain, the model can output different images.

Another limitation of CycleGANs is that it usually changes the background. For some tasks such as translating a horse to a zebra, we want to translate the individual objects only, without altering the background. The attention-guided CycleGANs (Alami Mejjati et al. 2018) method tackles this limitation by introducing unsupervised attention mechanisms. Specifically, two attention networks are adopted to locate the foreground and background areas. For the generator, the foreground object is created by applying the attention map to the translated image via an element-wise product, and the background image is created by applying the inverse attention map (i.e., $1 - s_{\text{attention}}$) to the input image via an element-wise product. Moreover, the discriminator is adapted to consider the foreground objects only, because the discriminator will treat the background image reserved from the input image as a fake image.

3.2 Unsupervised Domain Adaptation

Unsupervised domain adaptation addresses the problem of learning a classifier on the source domain with labels and applying it to a completely unlabeled target domain. GANs have achieved great success in application to unsupervised domain adaptation. In this section, we introduce three ways of applying GANs to unsupervised domain adaptation. The first method is to introduce a discriminator to judge whether the features come from the source or target domain, such that the features can be domain-invariant. The second method is based on image-to-image translation, by which we can generate target-domain samples with labels. The third method is based on multi-domain image generation. Because we can generate well-aligned image pairs, we can train an encoder to output similar features for the image pairs.

3.2.1 Domain Adversarial Training

Domain Adversarial Neural Network (DANN) (Ganin et al. 2016) has shown effectiveness in unsupervised domain adaptation and has been a basis for numerous domain adaptation methods (Tzeng et al. 2017; Long et al. 2018; Shu et al. 2018;

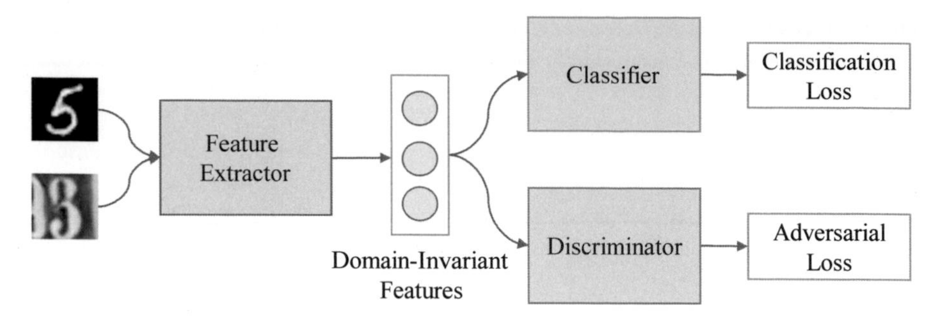

Fig. 3.4 Framework of DANN

Zhang et al. 2018). The idea of DANN is to generate domain-invariant features for the source and target domains by introducing a domain discriminator. Specifically, as shown in Fig. 3.4, DANN consists of a feature extractor, a domain discriminator, and a task classifier. The feature extractor and the domain discriminator are trained adversarially: the target of the discriminator is to distinguish whether the features generated by the extractor come from the source or the target domain, and the extractor tries to generate domain-invariant features.

Note that DANN is a concurrent work to GANs. The original DANN proposed a different method called gradient reversal layer to optimize the adversarial loss of the feature extractor. The gradient reversal layer multiplies the gradient from the discriminator by a certain negative constant during the backpropagation process. This technique corresponds to the minimax loss of GANs as described in Sect. 2.1.2. The minimax loss can cause a saturation problem (Goodfellow et al. 2014) during the early stage of training, and thus GANs replaced the minimax loss with the non-saturating loss in practice. As pointed out in Tzeng et al. (2017), the non-saturating loss also performs better than the minimax loss for domain adversarial training. Therefore, we introduce GANs training to optimize the adversarial loss of the feature extractor instead of the gradient reversal layer in this section.

Formally, let X_s and Y_s be the distributions of the input image x_s and label y_s from the source domain, and let X_t be the input distribution of the target domain. Suppose a classifier $F = H \circ G$ can be decomposed into a feature encoder G and an embedding classifier H. In contrast, a domain discriminator D maps the feature vector to the domain label (0, 1). The domain discriminator D and feature encoder G are trained adversarially, and the objective can be formalized as

$$
\begin{aligned}
&\min_{F,G} \max_{D} \mathcal{L}_{\text{cls}}(F) + \lambda \mathcal{L}_{\text{GAN}}(G, D), \\
&\mathcal{L}_{\text{cls}}(F) = -\mathbb{E}_{(x_s, y_s)}[y_s^\top \log F(x_s)], \\
&\mathcal{L}_{\text{GAN}}(G, D) = \mathbb{E}_{x_s}[\log D(G(x_s))] + \mathbb{E}_{x_t}[\log(1 - D(G(x_t)))],
\end{aligned}
\tag{3.6}
$$

where $\mathcal{L}_{\text{cls}}$ is the cross-entropy loss for classification, $\mathcal{L}_{\text{GAN}}$ is the adversarial loss, and λ controls the weight of the adversarial loss.

Next, we introduce four variants of DANN to improve the performance of DANN.

The Adversarial Discriminative Domain Adaptation (ADDA) (Tzeng et al. 2017) model improves the performance of DANN from two aspects. First, ADDA replaces the gradient reversal layer with the non-saturating loss of GANs (refer to Sect. 2.1.2), because the gradient reversal layer, which corresponds to the minimax loss of GANs, can cause the saturation problem (Goodfellow et al. 2014) during the early stage of training. Second, ADDA uses independent feature extractors for the source and target domains, which allows more domain-specific features to be learned. In practice, a pre-trained source extractor is used as the initialization for the target extractor; otherwise, the target extractor may learn a degenerate solution quickly.

It is well-known that conditional GANs (e.g., conditioning on the class labels) can generate higher-quality images than unconditional GANs, because conditioning on discriminative information enhances the ability of the discriminator, which in turn improves the performance of the generator. This is the intuition of the Conditional Adversarial Domain Adaptation (CDAN) (Long et al. 2018) model. CDAN makes the domain discriminator conditioned on the classifier predictions and the uncertainty of the classifier predictions, based on the fact that the predictions of the task classifier may convey discriminative information of whether the features come from the source or target domain, because the task classifier is trained using the source labels only.

The Virtual Adversarial Domain Adaptation (VADA) (Shu et al. 2018) model incorporates the cluster assumption (Grandvalet and Bengio 2005) constraint into the framework of DANN. The cluster assumption states that the decision boundaries should not cross the high-density regions, which has been extensively studied for semi-supervised learning (Grandvalet and Bengio 2005). Two techniques are adopted to constrain the cluster assumption: conditional entropy minimization (Grandvalet and Bengio 2005) and virtual adversarial training (Miyato et al. 2018). Furthermore, VADA introduces a training technique called Decision-Boundary Iterative Refinement Training (DIRT) to solely minimize the cluster assumption violation. This technique contains two phases: initializing with a VADA model and minimizing the cluster assumption violation in the target domain.

The key to the DANN model is to learn domain-invariant features. However, some domain-specific features such as corners and edges are useful for classifying images from different classes. Based on this observation, the Collaborative and Adversarial Network (CAN) (Zhang et al. 2018) model keeps domain-specific features in the lower layers of the neural network and encodes domain-invariant features in the higher layers. To keep the domain-specific features in the lower layers, several domain discriminators are incorporated on lower layers. In contrast to the feature extractor of DANN, the lower layers and the domain discriminators are both trained to distinguish whether the images come from the source or target domain. This is based on the intuitive idea that if the features extracted from the lower layers can be used to well distinguish whether the images come from the source or target domain, the features should convey sufficient domain-specific information.

3.2.2 *Using Image-to-Image Translation*

It is an intuitive way to use image-to-image translation for unsupervised domain adaptation. One simple method is that by translating a source-domain image into a corresponding target-domain image that belongs to the same class as the source image, we can collect a set of translated target-domain images with labels, and we can then train a classifier on the translated target-domain images. We introduce three models of applying image-to-image translation to unsupervised domain adaptation.

The PixelDA (Bousmalis et al. 2017) model uses cGANs (Mirza and Osindero 2014) to translate the source-domain images to target-domain images. The framework of PixelDA is similar to pix2pix (Isola et al. 2017). PixelDA consists of a generator, a discriminator, and a classifier. As described in Sect. 3.1.1, the use of cGANs for image-to-image translation requires the paired information of the training data. However, for unsupervised domain adaptation, the paired information is unavailable. The PixelDA model introduces two methods to constrain the semantic consistency between the source and translated images. The first method is to use the classification loss to force the generator to output images within the same class of source-domain images. The second method is to introduce a content-similarity loss that constrains the pixel-wise distance of the foreground objects between the source-domain and translated images.

Formally, let X_s and Y_s be the distributions of the input image x_s and label y_s from the source domain, and let X_t be the input distribution of the target domain. A generator G maps an image x_s from domain X to an image $G(x_s)$ from domain Y. A domain discriminator D maps the feature vector to the domain label (0, 1). A classifier F is trained on the source domain. The domain discriminator D and generator G are trained adversarially. The full objective of the PixelDA can be expressed as

$$\begin{aligned}
&\min_{F,G}\max_{D} \mathcal{L}_{\text{GAN}}(G, D) + \beta\mathcal{L}_{\text{cls}}(F, G) + \lambda\mathcal{L}_{\text{content}}(G),\\
&\mathcal{L}_{\text{GAN}}(G, D) = \mathbb{E}_{x_t}[\log D(x_t)] + \mathbb{E}_{x_s,z}[\log(1 - D(G(x_s, z)))],\\
&\mathcal{L}_{\text{cls}}(G, F) = -\mathbb{E}_{(x_s,y_s),z}[y_s^\top \log F(x_s) + y_s^\top \log F(G(x_s, z))],\\
&\mathcal{L}_{\text{content}}(G) = \mathbb{E}_{x_s,z}\left[\frac{1}{k}\,\|(x_s - G(x_s, z)) \circ m\|_2^2 - \frac{1}{k^2}\left((x_s - G(x_s, z))^\top m\right)^2\right],
\end{aligned} \tag{3.7}$$

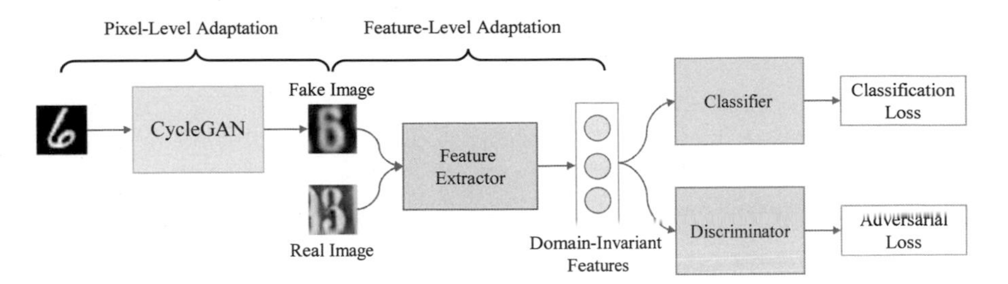

Fig. 3.5 Framework of CyCADA

where $\mathcal{L}_{\text{cls}}$ is the cross-entropy loss for classification, $\mathcal{L}_{\text{GAN}}$ is the adversarial loss, $\mathcal{L}_{\text{content}}$ is the content-similarity loss using PMSE (Eigen et al. 2014), m is a binary mask, and (β, λ) controls the relative importance of the losses.

However, PixelDA only works for simple datasets and small domain shifts, which limits the applicability of PixelDA. In Bousmalis et al. (2017), PixelDA is evaluated only on several simple datasets, such as MNIST, MNIST-M, and synthetic 3D objects.

As described in Sect. 3.1.2, the CycleGANs (Zhu et al. 2017) model can perform image-to-image translation in the absence of the paired images. The CyCADA (Hoffman et al. 2018) model uses CycleGANs for unsupervised domain adaptation. The framework of CyCADA is shown in Fig. 3.5. CyCADA considers both the pixel-level adaptation and feature-level adaptation. The pixel-level adaptation is achieved with the use of CycleGANs to translate the source-domain images into target-domain images and training a task classifier on the translated images. The feature-level adaptation is achieved by the use of domain adversarial training in Sect. 3.2.1, that is, training a domain discriminator to distinguish whether the features come from the source or target domain. Furthermore, CyCADA uses a content loss to constrain the semantic consistency before and after image translation. Specifically, a pre-trained source-domain classifier is used, and CyCADA ensures that the images should be classified the same way before and after translation according to the pre-trained classifier.

The SBADA-GANs (Russo et al. 2018) model adapts CycleGANs (Zhu et al. 2017) to unsupervised domain adaptation in another way. Compared with CyCADA, SBADA-GANs focus more on class-level information. CyCADA keeps the cycle-consistency loss of CycleGANs as a pixel-level content loss and adopts the L1 norm to measure the distance between the source and reconstructed images, whereas SBADA-GANs adapt the cycle-consistency loss to a class-level consistency loss. Specifically, the reconstructed images should be correctly classified by a classifier pre-trained on the source domain. Moreover, SBADA-GANs also use a classifier to guide the generator to output images with class information reserved. The classifier is trained on the translated images with the labels of the input images, and the classification loss is backpropagated to the generator.

3.2.3 *Using RCGANs*

As described in Sect. 2.4, RCGANs can generate well-aligned image pairs from different domains in the absence of paired samples. Because the aligned images belong to the same class, we can enforce a feature extractor to output similar features for the aligned images. Specifically, as Fig. 3.6 shows, given a pair of inputs with identical latent variables but different domain variables, the generator of RCGANs can generate aligned images that belong to the same class. We can add a regularizer to the last hidden layer of the discriminator, which forces the last hidden layer to output similar feature representations for the aligned images. A classifier is attached to the last hidden layer of the discriminator and is trained with the images from the source domain with labels. This classifier can be used for the target domain because the features of the target domain are similar to those of the source domain. The

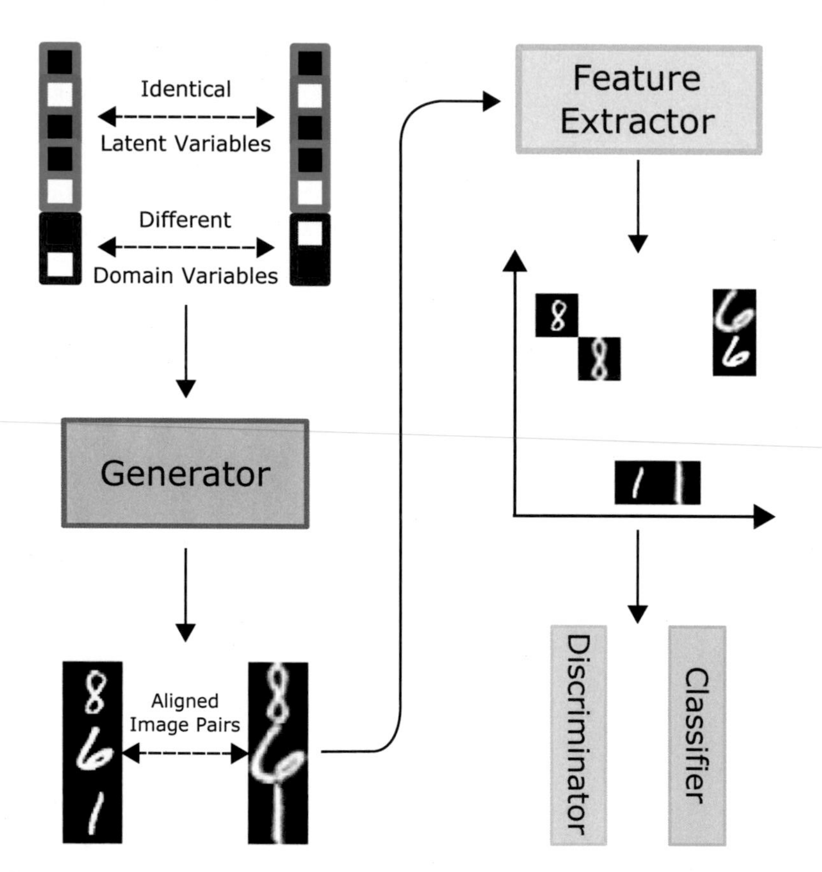

Fig. 3.6 Left: Given a pair of inputs with identical latent variables but different domain variables, the generator maps them to a pair of aligned images. Right: Using the aligned images as input, the feature extractor extracts similar feature representations for the aligned images. The discriminator and classifier share the features extracted by the feature extractor

objective can be formulated as

$$\min_{G,C} \max_{D} V(G, D, C) = \mathcal{L}_{\text{GAN}}(G, D) + \lambda \mathcal{L}_{\text{reg}}(G) + \beta \mathcal{L}_{\text{reg}}(D) + \gamma \mathcal{L}_{\text{cls}}(C), \tag{3.8}$$

where the scalars λ, β, and γ are used to adjust the weights of the regularization terms and the classifier and $\mathcal{L}_{\text{cls}}(C)$ is a typical cross-entropy loss.

Note that this approach differs from DANN (Ganin et al. 2016), which tries to minimize the difference between the overall distribution of the source and target domains. In contrast, the minimization of RCGANs is among the samples that belong to the same category because it only penalizes the distances between the pairs of corresponding images that belong to the same category.

3.2.3.1 Experiments

We evaluate the proposed method on MNIST and USPS datasets; one is used as the source domain and the other as the target domain. We set $\lambda = 0.1$, $\beta = 0.004$, and $\gamma = 1.0$ found by grid search. Following the experimental protocol in Liu and Tuzel (2016) and Tzeng et al. (2017), we randomly sample 2000 images from MNIST and 1800 images from USPS.

We conduct two comparison experiments between RCGANs and the baseline methods, including DANN (Ganin et al. 2016), ADDA (Tzeng et al. 2017), and CoGAN (Liu and Tuzel 2016). One is to evaluate the classification accuracy directly on the sampled images of the target domain, which was adopted by Liu and Tuzel (2016) and Tzeng et al. (2017). To further evaluate the generalization error, we further evaluate the classification accuracy on the standard test sets of the target domain.

The results are presented in Table 3.1. The reported accuracies are averaged over 10 trials with different random samplings. For the evaluation on the standard test set, RCGANs significantly outperform all of the baseline methods, especially for the task of USPS to MNIST. This shows that RCGANs have a smaller generalization error than the baseline methods. For the evaluation on the sampled set, RCGANs outperform all of the baseline methods for the task of MNIST to USPS and achieve comparable performance for the task of USPS to MNIST. The generated pairs of MNIST and USPS digits are shown in Fig. 3.7.

We find that the regularizer in the generator also plays an important role when applying RCGANs to domain adaptation. Compared with the technique of weight sharing (Liu and Tuzel 2016), regularizing the high-level semantics in the generator is more flexible and effective, as demonstrated by experiments. If the noise input z is not conditioned by the domain variables in Fig. 2.18a, the output of the generator's first layer will be the same for different domains. This turns out to be weight sharing. The comparison results are presented in Table 3.2, where $\lambda = 0.0$ represents the model without the regularizer or weight sharing. From Table 3.2, we have three main

Table 3.1 Accuracy results for unsupervised domain adaptation. Top section presents the classification accuracy evaluated on the sampled set of the target domain. Bottom section presents the classification accuracy evaluated on the standard test set of the target domain. The reported accuracies are averaged over 10 trials with different random samplings

Method	MNIST→USPS	USPS→MNIST
	Evaluated on the sampled set	
DANN	0.771	0.730
ADDA	0.894 ± 0.002	**0.901** ± 0.008
CoGANs	0.912 ± 0.008	0.891 ± 0.008
RCGANs (Ours)	**0.931** ± 0.007	0.895 ± 0.009
	Evaluated on the test set	
ADDA	0.836 ± 0.035	0.849 ± 0.058
CoGANs	0.882 ± 0.018	0.822 ± 0.081
RCGANs (Ours)	**0.901** ± 0.009	**0.888** ± 0.015

Fig. 3.7 Generated pairs of MNIST and USPS digits

Table 3.2 Comparison results between the regularizer in the generator and weight sharing. λ is the weight of the regularizer in Equation 2.18. Classification accuracy is evaluated on the standard test set of the target domain

Method	MNIST→USPS	USPS→MNIST
$\lambda = 0.0$	0.882 ± 0.014	0.857 ± 0.017
Weight Sharing	0.889 ± 0.011	0.876 ± 0.018
Regularizer ($\lambda = 0.1$)	**0.901** ± 0.009	**0.888** ± 0.015

observations. First, both the regularizer and weight sharing can improve accuracy when compared with the baseline model (i.e., $\lambda = 0.0$). Second, the addition of a regularizer is more effective than weight sharing, possibly because the regularizer is more flexible than weight sharing, which enforces the model to output identical values. Third, the model with $\lambda = 0.0$ also outperforms ADDA and CoGAN, which shows the effectiveness of our proposed framework.

3.3 GANs for Security

The word *Deepfake* is a portmanteau of the words "deep learning" and "fake." Deepfakes are synthetic media content generated by techniques from artificial intelligence in which a synthetic person in an image or video looks authentic in the eyes of humans. Deepfakes have been applied to many areas (Mirsky and Lee 2020), such as movies, education, and entertainment. However, despite the positive applications, deepfakes have gained widespread attention for their unethical use in pornographic videos and fake politician videos.[1]

The most influential deepfake techniques are face reenactment and face swapping. Face reenactment is the task of using the expression and pose of a source person to drive the expression and pose of a target person. Face swapping is the task of replacing the face of a target person with the face of a source person, preserving the face identity of the source person while keeping other things unchanged.

Earlier works Blanz et al. (2004), Bitouk et al. (2008), and Thies et al. (2016) used 3D graphics techniques to perform face reenactment and swapping. Recently, GANs have been used to generate deepfakes because of their powerful ability to generate fake images. In this section, we focus on the methods by which GANs are applied to face reenactment and swapping. We first introduce four models that use GANs to generate deepfake images or videos and then introduce four methods to detect deepfakes.

The Deep Video Portraits (DVP) (Kim et al. 2018) model is the first that can transfer the full head position, face expression, eye gaze, and eye blinking from a source person to a video of a target person. The idea is to render synthetic images of the target person first and then to utilize the techniques of image-to-image translation (Sect. 3.1) to translate the synthetic images into photo-realistic target videos. The DVP model needs both source and target videos as input. The processes of DVP consist of four steps: (1) extract a low-dimensional parametric representation of both the source and target videos based on a parametric head model (Blanz and Vetter 1999); (2) transfer the parameters of the source video that correspond to the head pose, face expression, and eyes into the parameter space of the target video; (3) render synthetic images under the parameters in the second step using the hardware rasterization; and (4) use the techniques based on GANs to translate the synthetic images into photo-realistic target videos. For the fourth step, the network architecture is similar to pix2pix (Isola et al. 2017). The inputs of the generator are full frames of the synthetic images, and the generator translates the synthetic images into photo-realistic target videos. For the discriminator, DVP follows the pix2pix model to use a patch-based network as described in Sect. 3.1.1.

The Recycle-GANs (Bansal et al. 2018) model adapts the CycleGANs (Zhu et al. 2017) model for video-to-video translation. Like CycleGAN, Recycle-GANs adopt the adversarial loss and the cycle-consistency loss for static image translation of

[1] https://en.wikipedia.org/wiki/Deepfake

each frame. To consider the temporal information of videos, the recycle-consistency loss is proposed, which is based on a recurrent temporal predictor that predicts future frames based on the previous frames. The recycle-consistency loss consists of three steps: (1) the generator translates several frames of the source video to the style of the target video, (2) the recurrent temporal predictor predicts the next frame of the translated images, and (3) the reverse generator translates the predicted frame back to the style of the source video and this translated frame should be similar to the ground truth frame of the source video.

DVP and Recycle-GANs both require a video of the target person. The photoreal avatar GANs (paGANs) (Nagano et al. 2018) model can perform face reenactment using only one image of a target person. What's more, paGANs can use only one network for all different target persons, whereas DVP requires different networks trained for different target persons. The processes of paGANs consist of four steps: (1) fit a 3D morphable model (Thies et al. 2016) to the source and target images to obtain the facial parameters; (2) blend the parameters of the target identity and source expression; (3) construct the input of the network, including the target image, the depth image of the blended parameters, the normal direction image of the blended parameters, a masked image for eye gaze, and a synthetic image of the blended parameters; and (4) use the pix2pix (Isola et al. 2017) model to translate the input into a photo-realistic face image. Moreover, paGANs can perform real-time face reenactment on a mobile device by pre-computing several key expression textures.

The Face Swapping GANs (FSGANs) (Nirkin et al. 2019) model is designed for face swapping. Because face swapping only requires replacement of the face while keeping the hairstyle of the source person, FSGANs use a segmentation network to predict the face and hair masks. The processes of FSGANs consist of four steps: (1) extract the facial landmarks of both source and target persons; (2) use one pix2pixHD (Wang et al. 2017) model to translate the target person into the source person while retaining the pose and expression of the target person; (3) extract the face region using the segmentation network and paste this face to the target image; (4) construct the input for another pix2pixHD model, including the real target image, the combined image in the third step, and the segmentation mask of the target image; and (5) use the second pix2pixHD model to translate the input into a photo-realistic face swapped image.

Next, we introduce four methods to detect the deepfakes generated by the above techniques. In general, deepfake detection methods formulate this problem as a binary classification problem and train a classifier to distinguish fake images or videos from authentic cases.

A convolutional recurrent neural network (CNN-RNN) was proposed for deepfake video detection by Güera and Delp (2018), which uses a convolutional neural network (CNN) to extract frame-level features and a recurrent neural network (RNN) to extract sequence-level features. Specifically, the InceptionV3 (Szegedy et al. 2016) model pre-trained on the ImageNet dataset is used to extract features for each frame, and the Long Short Term Memory (LSTM) network is used for sequence processing using the frame-level features as the input.

The eye-blinking detection model (Li et al. 2018) exposes deepfake videos by detecting eye blinking. It is based on the fact that a healthy adult human generally blinks every 2 to 10 seconds. However, this is not the case for many deepfake videos because most images used to train GANs are of people with their eyes open. This method first extracts a surrounding rectangular region of the eyes in each frame and then feeds the sequence of the extracted eye regions as the input to a long-term recurrent convolutional neural networks (LRCN) model (Donahue et al. 2015) that predicts whether the eye is open or closed for each frame.

The Leader Detection model (Agarwal et al. 2019) is designed for deepfake detection of a specific person, such as a national leader. It is based on the observation that different individuals exhibit relatively distinct patterns of facial and head movements when speaking. The idea of the Leader Detection model is to analyze the speaking pattern of a specific person and then train a classifier to distinguish this pattern from the patterns in the fake videos. The speaking pattern is analyzed by first extracting 20 facial/head features, such as head pose, and then computing the Pearson correlation between these 20 features. Using the 190 pairs of features as the input, a one-class support vector machine (SVM) (Schölkopf et al. 1999) is trained that requires only authentic videos of a person.

The FakeCatcher (Ciftci and Demir 2020) model finds that the blind use of deep learning is not effective in catching deepfakes and uses the biological signals hidden in portrait videos as an implicit feature. FakeCatcher first extracts six biological signals by combining G-PPG (Zhao et al. 2018) and C-PPG (de Haan and Jeanne 2013) on the left cheek, right cheek (Feng et al. 2015), and mid-region (Tulyakov et al. 2016). FakeCatcher then performs statistical transformations on these signals, such as log scale and power spectral density. These statistical features are used as the input for the classification models. Two classification models are trained, including an SVM to achieve interpretable features and a CNN to achieve greater accuracy.

Bibliography

Agarwal S, Farid H, Gu Y, He M, Nagano K, Li H (2019) Protecting world leaders against deep fakes. In: Computer vision and pattern recognition (CVPR) workshops, pp 38–45

Alami Mejjati Y, Richardt C, Tompkin J, Cosker D, Kim KI (2018) Unsupervised attention-guided image-to-image translation. Adv Neural Inf Process Syst 31:3693–3703

Almahairi A, Rajeswar S, Sordoni A, Bachman P, Courville A (2018) Augmented CycleGAN: learning many-to-many mappings from unpaired data. In: International conference on machine learning (ICML)

Anoosheh A, Agustsson E, Timofte R, Gool LV (2018) ComboGAN: unrestrained scalability for image domain translation. In: Computer vision and pattern recognition (CVPR) workshops, pp 783–790

Bansal A, Ma S, Ramanan D, Sheikh Y (2018) Recycle-GAN: unsupervised video retargeting. In: European conference on computer vision (ECCV)

Benaim S, Wolf L (2017) One-sided unsupervised domain mapping. In: Advances in neural information processing systems (NeurIPS), pp 752–762

Bitouk D, Kumar N, Dhillon S, Belhumeur P, Nayar SK (2008) Face swapping: automatically replacing faces in photographs. ACM Trans Graph 27:39
Blanz V, Vetter T (1999) A morphable model for the synthesis of 3D faces. In: Annual conference on computer graphics and interactive techniques (SIGGRAPH), pp 187–194
Blanz V, Scherbaum K, Vetter T, Seidel H-P (2004) Exchanging faces in images. Comput Graph Forum 23:669–676
Bousmalis K, Silberman N, Dohan D, Erhan D , Krishnan D (2017) Unsupervised pixel-level domain adaptation with generative adversarial networks. In: Computer vision and pattern recognition (CVPR), pp 3722–3731
Choi Y, Choi M, Kim M, Ha J-W, Kim S, Choo J (2018) StarGAN: unified generative adversarial networks for multi-domain image-to-image translation. In: Computer vision and pattern recognition (CVPR), pp 8789–8797
Ciftci UA, Demir I (2020) FakeCatcher: detection of synthetic portrait videos using biological signals. IEEE Trans Pattern Anal Mach Intell, 10.1109/TPAMI.2020.3009287
de Haan G, Jeanne V (2013) Robust pulse rate from chrominance-based rPPG. IEEE Trans Biomed Eng 60:2878–2886
Denton E, Chintala S, Szlam A, Fergus R (2015) Deep generative image models using a Laplacian pyramid of adversarial networks. In: Advances in neural information processing systems (NeurIPS), pp 1486–1494
Donahue J, Hendricks LA, Rohrbach M, Venugopalan S, Guadarrama S, Saenko K, Darrell T (2015) Long-term recurrent convolutional networks for visual recognition and description. In: Computer vision and pattern recognition (CVPR), pp 2625–2634
Eigen D, Puhrsch C, Fergus R (2014) Depth map prediction from a single image using a multi-scale deep network. In: Advances in neural information processing systems (NeurIPS), pp 2366–2374
Feng L, Po L-M, Xu X, Li Y, Ma R (2015) Motion-resistant remote imaging photoplethysmography based on the optical properties of skin. IEEE Trans Biomed Eng 25:879–891
Ganin Y, Ustinova E, Ajakan H, Germain P, Larochelle H, Laviolette F, Marchand M, Lempitsky V (2016) Domain-adversarial training of neural networks. J Mach Learn Res 17(1):2096–2030
Goodfellow I, Pouget-Abadie J, Mirza M, Xu B, Warde-Farley D, Ozair S, Courville A, Bengio Y (2014) Generative adversarial nets. In: Advances in neural information processing systems (NeurIPS), pp 2672–2680
Grandvalet Y, Bengio Y (2005) Semi-supervised learning by entropy minimization. In: Conference on neural information processing systems (NeurIPS), pp 529–536
Güera D, Delp EJ (2018) Deepfake video detection using recurrent neural networks. In: International conference on advanced video and signal based surveillance (AVSS)
Hoffman J, Tzeng E, Park T, Zhu J-Y, Isola P, Saenko K, Efros AA, Darrell T (2018) CyCADA: cycle-consistent adversarial domain adaptation. In: International conference on machine learning (ICML), pp 1989–1998
Huang X, Li Y, Poursaeed O, Hopcroft J, Belongie S (2017) Stacked generative adversarial networks. In: Computer vision and pattern recognition (CVPR), pp 5077–5086
Isola P, Zhu J-Y, Zhou T, Efros AA (2017) Image-to-image translation with conditional adversarial networks. In: Computer vision and pattern recognition (CVPR), pp 5967–5976
Johnson J, Alahi A, Fei-Fei L (2016) Perceptual losses for real-time style transfer and super-resolution. In: European conference on computer vision (ECCV)
Kim H, Garrido P, Tewari A, Xu W, Thies J, Nießner M, Pérez P, Richardt C, Zollhöfer M, Theobalt C (2018) Deep video portraits. ACM Trans Graph 37:1–14
Lee H-Y, Tseng H-Y, Huang J-B, Singh MK, Yang M-H (2018) Diverse image-to-image translation via disentangled representations. In: European conference on computer vision (ECCV)
Li C, Liu H, Chen C, Pu Y, Chen L, Henao R, Carin L (2017) ALICE: towards understanding adversarial learning for joint distribution matching. In: Advances in neural information processing systems (NeurIPS), pp 5495–5503
Li Y, Chang M-C, Lyu S (2018) In Ictu Oculi: exposing AI created fake videos by detecting eye blinking. In: International workshop on information forensics and security (WIFS)

Liu M-Y, Tuzel O (2016) Coupled generative adversarial networks. In: Advances in neural information processing systems (NeurIPS)
Liu M-Y, Breuel T, Kautz J (2017) Unsupervised image-to-image translation networks. In: Advances in neural information processing systems (NeurIPS), pp 700–708
Long M, Cao Z, Wang J, Jordan MI (2018) Conditional adversarial domain adaptation. In: Conference on neural information processing systems (NeurIPS), pp 1645–1655
Ma S, Fu J, Chen CW, Mei T (2018) DA-GAN: instance level image translation by deep attention generative adversarial networks. In: Computer vision and pattern recognition (CVPR), pp 5657–5666
Mirsky Y, Lee W (2020) The creation and detection of deepfakes: a survey. arXiv:2004.11138
Mirza M, Osindero S (2014) Conditional generative adversarial nets. arXiv:1411.1784
Miyato T, Maeda S, Koyama M, Ishii S (2018) Virtual adversarial training: a regularization method for supervised and semi-supervised learning. IEEE Trans Pattern Anal Mach Intell 41:1979–1993
Nagano K, Seo J, Xing J, Wei L, Li Z, Saito S, Agarwal A, Fursund J, Li H (2018) paGAN: real-time avatars using dynamic textures. ACM Trans Graph 37:1–12
Nirkin Y, Keller Y, Hassner T (2019) FSGAN: subject agnostic face swapping and reenactment. In: International conference on computer vision (ICCV), pp 7184–7193
Park T, Liu M-Y, Wang T-C, Zhu J-Y (2019) Semantic image synthesis with spatially-adaptive normalization. In: Computer vision and pattern recognition (CVPR), pp 2337–2346
Park T, Efros AA, Zhang R, Zhu J-Y (2020) Contrastive learning for unpaired image-to-image translation. In: European conference on computer vision (ECCV)
Ronneberger O, Fischer P, Brox T (2015) U-Net: convolutional networks for biomedical image segmentation. In: Medical image computing and computer-assisted intervention (MICCAI), pp 234–241
Russo P, Carlucci FM, Tommasi T, Caputo B (2018) From source to target and back: symmetric bi-directional adaptive GAN. In: Computer vision and pattern recognition (CVPR), pp 8099–8108
Schölkopf B, Platt JC, Shawe-Taylor JC, Smola AJ, Williamson RC (1999) Estimating the support of a high-dimensional distribution. Neural Comput 13:1443–1471
Shu R, Bui H, Narui H, Ermon S (2018) A DIRT-T approach to unsupervised domain adaptation. In: International conference on learning representations (ICLR)
Szegedy C, Vanhoucke V, Ioffe S, Shlens J, Wojna Z (2016) Rethinking the inception architecture for computer vision. In: Computer vision and pattern recognition (CVPR), pp 2818–2826
Thies J, Zollhöfer M, Stamminger M, Theobalt C, Niessner M (2016) Face2Face: real-time face capture and reenactment of RGB videos. In: Computer vision and pattern recognition (CVPR)
Tulyakov S, Alameda-Pineda X, Ricci E, Yin L, Cohn JF, Sebe N (2016) Self-adaptive matrix completion for heart rate estimation from face videos under realistic conditions. In: Computer vision and pattern recognition (CVPR)
Tzeng E, Hoffman J, Saenko K, Darrell T (2017) Adversarial discriminative domain adaptation. In: Computer vision and pattern recognition (CVPR)
van den Oord A, Li Y, Vinyals O (2018) Representation learning with contrastive predictive coding. arXiv:1807.03748
Wang T-C, Liu M-Y, Zhu J-Y, Tao A, Kautz J, Catanzaro B (2017) High-resolution image synthesis and semantic manipulation with conditional GANs. arXiv:1711.11585
Wang C, Xu C, Wang C, Tao D (2018) Perceptual adversarial networks for image-to-image transformation. IEEE Trans Image Process 27:4066–4079
Zhang H, Xu T, Li H, Zhang S, Huang X, Wang X, Metaxas D (2016) StackGAN: text to photo-realistic image synthesis with stacked generative adversarial networks. arXiv:1612.03242
Zhang W, Ouyang W, Li W, Xu D (2018) Collaborative and adversarial network for unsupervised domain adaptation. In: Computer vision and pattern recognition (CVPR)
Zhao C, Lin C-L, Chen W, Li Z (2018) A novel framework for remote photoplethysmography pulse extraction on compressed videos. In: Computer vision and pattern recognition (CVPR) workshops, pp 1299–1308

Zhu J-Y, Park T, Isola P, Efros AA (2017a) Unpaired image-to-image translation using cycle-consistent adversarial networks. In: International conference on computer vision (ICCV)

Zhu J-Y, Zhang R, Pathak D, Darrell T, Efros AA, Wang O, Shechtman E (2017b) Toward multimodal image-to-image translation. In: Advances in neural information processing systems (NeurIPS), pp 465–476

Chapter 4
Conclusions

4.1 Contributions

GAN is a powerful model for image generation, which has been a fundamental basis for numerous computer vision tasks. In this book, we have described several models for image generation and multi-domain image generation. For image generation, we introduced the Least Squares Generative Adversarial Networks (LSGANs), which use the least squares loss for both the generator and discriminator, instead of using the cross-entropy loss. The idea of how to overcome the vanishing gradients problem during GANs learning was explained via both intuitive examples and theoretical analysis. The results showed that LSGANs not only generate higher-quality images but also has a more stable performance than regular GANs. For multi-domain image generation, we introduced the Regularized Conditional Generative Adversarial Networks (RCGANs), which use the conditional GANs and two regularizers to force the model to encode the domain information in the conditioned domain variables. One regularizer is added to the first layer of the generator to guide the generator to decode similar high-level semantics. The other is added to the last hidden layer of the discriminator to force the discriminator to output similar losses for the corresponding images. We also introduced a method of applying RCGAN to unsupervised domain adaptation.

We also introduced three interesting applications of GANs, including image-to-image translation, unsupervised domain adaptation, and deepfakes. For image-to-image translation, we introduced two famous models, pix2pix and CycleGANs, that are widely used in computer vision and graphics tasks. The pix2pix model is designed for the scenario in which the paired information is available, whereas the CycleGANs model is designed for scenarios in which the paired information is unavailable. For unsupervised domain adaptation, we introduced domain adversarial training, which uses a discriminator to extract domain-invariant features for the source and target domains. For deepfakes, we introduced four interesting GAN-based models that can perform high-quality face reenactment and face swapping.

X. Mao, Q. Li, *Generative Adversarial Networks for Image Generation*,
https://doi.org/10.1007/978-981-33-6048-8_4

4.2 Future Research

In terms of improving the generated image quality, numerous GAN models have evolved from training on MNIST (Goodfellow et al. 2014; Radford et al. 2015) to training on ImageNet (Zhang et al. 2018; Brock et al. 2018) in terms of network architectures (Radford et al. 2015; Denton et al. 2015), loss functions (Mao and Li 2018), self-attention (Zhang et al. 2018), and sampling methods (Brock et al. 2018). Although GAN is very powerful for image generation, challenges remain. Improvement of the generated image quality remains a hot topic in the community. Currently, generating photorealistic images is still limited to some simple datasets, in which the objects are usually centered with small margins. For example, the CELEBA-HQ dataset is created by centering the faces and cropping the margins. For complex datasets that are more dynamic and structured, such as human bodies and scene images (Brock et al. 2018), the generated image quality is still far from photorealistic, and people can distinguish the fake images easily. Therefore, learning the global structure of the complex images may be a key to improve the performance of GAN for complex datasets that is worthy of study.

Regularizing the Lipschitz constant of the discriminator has been proved the effectiveness of improving the training stability of GAN. Several methods have been proposed to regularize the Lipschitz constant by weight decay (Qi 2017), gradient penalty (Gulrajani et al. 2017), and spectral normalization (Miyato et al. 2018). Therefore, we believe that exploring more effective ways of regularizing the Lipschitz constant of the discriminator is another promising way to improve the performance of GAN.

A recent state-of-the-art model named BigGAN (Brock et al. 2018) revealed that increasing the training batch size can significantly improve the performance of GANs. It also noted that sampling the noise vector from a truncated normal distribution is a trick that can be used to improve the generated image quality. Therefore, designing more effective and memory-efficient network architectures and designing new sampling methods are worthy of further investigation.

Bibliography

Brock A, Donahue J, Simonyan K (2018) Large scale GAN training for high fidelity natural image synthesis. arXiv:1809.11096

Denton E, Chintala S, Szlam A, Fergus R (2015) Deep generative image models using a Laplacian pyramid of adversarial networks. In: Advances in neural information processing systems (NeurIPS), pp 1486–1494

Goodfellow I, Pouget-Abadie J, Mirza M, Xu B, Warde-Farley D, Ozair S, Courville A, Bengio Y (2014) Generative adversarial nets. In: Advances in neural information processing systems (NeurIPS), pp 2672–2680

Gulrajani I, Ahmed F, Arjovsky M, Dumoulin V, Courville A (2017) Improved training of Wasserstein GANs. In: Advances in neural information processing systems (NeurIPS), pp 5767–5777

Mao X, Li Q (2018) Unpaired multi-domain image generation via regularized conditional GANs. In: International joint conference on artificial intelligence (IJCAI)

Miyato T, Kataoka T, Koyama M, Yoshida Y (2018) Spectral normalization for generative adversarial networks. arXiv:1802.05957

Qi G-J (2017) Loss-sensitive generative adversarial networks on Lipschitz densities. arXiv:1701.06264

Radford A, Metz L, Chintala S (2015) Unsupervised representation learning with deep convolutional generative adversarial networks. arXiv:1511.06434

Zhang H, Goodfellow I, Metaxas D, Odena A (2018) Self-attention generative adversarial networks. arXiv:1805.08318

Printed in the United States
By Bookmasters